T0357267

The
Grieving
BODY

Also by Dr. Mary-Frances O'Connor

The Grieving Brain

The
Grieving
BODY

How the Stress of Loss Can Be
an Opportunity for Healing

MARY-FRANCES O'CONNOR, PhD

HarperOne
An Imprint of HarperCollins*Publishers*

This book contains advice and information relating to health care. It should be used to supplement rather than replace the advice of your doctor or another trained health professional. If you know or suspect you have a health problem, it is recommended that you seek your physician's advice before embarking on any medical program or treatment. All efforts have been made to assure the accuracy of the information contained in this book as of the date of publication. This publisher and the author disclaim liability for any medical outcomes that may occur as a result of applying the methods suggested in this book.

FIRST EDITION

Designed by Janet Evans-Scanlon

Library of Congress Cataloging-in-Publication Data has been applied for.

ISBN 978-0-06-333890-6
ISBN 978-0-06-345047-9 (ANZ)

24 25 26 27 28 LBC 5 4 3 2 1

To Rick,
for teaching me there is more to love than words

Contents

The
Grieving
BODY

Introduction

When a Part of "Us" Has Been Cut Away

I magine you are cycling down a city street lined with beautiful trees, enjoying the fresh air and slight breeze. You see that you are approaching an intersection, and you automatically reach out to squeeze your hand brakes. But there are no brakes there, only the bare handlebar. Your heart rate speeds up; your muscles tense. The intersection is getting closer—you need to stop. You clutch at the bar. And then you remember that this beach cruiser bicycle has coaster brakes, engaged by pedaling backward. Your heart may continue to pound for a little while as you travel on. Many of us would find the ride stressful, since the habitual way of riding a bike is no longer easy, no longer relaxing.

The key to this metaphor is that it is the *absence* of something, something built into the way we naturally operate in the world, that causes our body to react, our heart rate to accelerate, our blood pressure to surge. This whole situation happens in a brief moment, and may even happen largely in our subconscious. We figure out how to manage the situation. But even if we do not note it consciously as stressful, our body has still reacted to the absence of the brakes. You may never have considered this, but this is similar to how our body reacts to the *absence* of a loved one after they die. Our loved one is built into every action we take and how we function in the world. Their absence causes our body to react, because we need them in a basic, survival kind of way. Our body's stress response to their absence, over and over again, happens even if we dismiss the moment as unimportant. The missing hand brakes is a simplistic metaphor, but imagine how many times after a loss we might reach out for our one-and-only, by picking up the phone to call them, or even just realizing they are not in the room. When they are not there, our grieving body reacts.

When a loved one dies, it is not just our brain that responds. Our reaction to loss is not only in our thoughts, our emotions, our mind. The response to the death of a loved one is a physiological one as well, reverberating through our body. Bereaved people show increased heart rate, blood pressure, stress hormones, inflammation. Without our spouse, or our child, or our sibling, the world feels all wrong. Even when we cannot put our finger on what is missing from the situation (maybe it's missing the shoes kicked off and left lying in the hallway, or the whistle of the teakettle from the next room), our body still reacts to their absence.

When we form a bond with a loved one, we form an unconscious dynamic system, in addition to the conscious one we typically think of. This relationship between two bodies has many responses without either person's conscious awareness. When my partner hugs me, my heart rate decelerates. When I go to sleep at night, I respond to the warmth, smell, and sounds of him next to me, helping to cue the release and inhibition of various neurochemicals that allow me to slip into that state of slumber. Every time I leave on a trip, my departure from him resounds through my body in ways I am not aware of but that are nonetheless physiologically real. And our reunion leads to more changes in my body, experienced as comfort and familiarity upon my return, as my hormones return to their optimal set point. My body, in addition to my mind, is in a relationship with him. The death of a loved one reveals this system, which otherwise functions as an invisible force, a magnetic attachment bond that dances electron around nucleus without being seen by the naked eye. My body simply responds to my loved one's absence, unconsciously.

Human beings attend to the things we consciously experience. But there is more to love than words, and there is more to grief than the sadness we feel. Previously, I have written about grief and grieving through the neuroscientific lens. Much (although not all) of the mind is governed by thoughts, expressing what we are feeling and thinking, explaining our motivations and intentions, planning our activities. Not all of our grief can be expressed in words, and thoughts and insights do not always control our feelings or emotional reactions. *The Grieving Body* goes beyond the conscious experience to reveal the physiological impacts of grieving, what happens when "you" and "me" turn into "us" through bonding, and then "us" has a piece cut away. When this amputation happens, we do not return to the "me" of before, because the absence leaves a hole that could not have existed before we ever knew love.

Attachment relationships regulate our physiology, affecting our hormones and neurochemistry, our cardiovascular and immune systems. This dynamic relationship system governs our reactions and interactions. Our loved one is a huge resource in the ledger of what psychologist Lisa Feldman Barrett calls our "body budget," the constant attempt to keep our incoming and outgoing energy in balance.[1] Because our brain relies on our loved one to help us calm down or to get us motivated, our loved one usually helps to keep our body in the normal range of energy where we thrive. Consequently, during grieving, our body attempts to compensate for the hole that has been left and tries to reregulate. Attempting to reregulate during grieving looks like fatigue or brain fog or restlessness or being more susceptible to the flu. Figuring out how to regain the equilibrium in our physiological systems, doing everything without our loved one, is a (largely) unconscious process in grieving.

The book you are about to read tells the story of how our psychology, our nervous system, and our immune system are intimately linked, and how our bonds with our loved ones regulate these systems. I am a professor in the Psychology Department at the University of Arizona, and my research on inflammatory markers and cardiovascular function during bereavement, along with the research of many colleagues around the world, has advanced a mechanistic understanding of the broken-heart phenomenon. But in addition to my scientific research, I have been learning through personal life experience, through my own grief. Grieving, as I wrote about in my first book, *The Grieving Brain*, can be thought of as a form of learning. Grieving can also be an opportunity for healing. The opportunity arises if we can listen to the messages of stress in our body, those internal sensations and emotions. We can slowly begin to respond to them in a healthy and compassionate way, restoring a full life that takes our body's needs into account.

When my mother died in December 1999, our family had been anticipating it for years. When I was thirteen years old, she was diagnosed with stage IV breast cancer, followed by a mastectomy and a year of weekly chemotherapy, which led to days of her never changing out of her bathrobe. She then spent many years in remission, followed by a metastatic recurrence, and then a shorter time until the next metastasis, and so on. But all of this medical management was nothing compared to the painful management of our relationship, as I entered high school and then moved over a thousand miles away to college. I explored the world, motivated by that outward-focused stage of my life, trying to create a little space between us while feeling her emotional dependence on me and her heartbreaking grief in my absence. Navigating this change in my closeness with her, navigating my enduring love for her while I developed greater closeness with romantic attachments and friends, felt impossible.

When she died, I was relieved to no longer be torn between seeking my way in the world and returning repeatedly to Montana, where I grew up. The expectation that I be present at her bedside, or wrestle with guilt when I was not, finally ceased. But of course, our loved ones do not live only in the outer, physical world. They also live on in the virtual world of our mind, encoded in the neural connections of our brain, and my internal relationship with my mother continued as well. Her approval still mattered to me even though she was no longer there in person. Even after our Sunday phone calls ceased, I stressed about not meeting expectations she would have had. When others I cared for experienced depression or anxiety, I continued to see it as my job to fix it, just as I had tried to do for her. Our relationship continued to shape me long after her death, and my anxiety and strong emotional reactivity, which I attributed to interactions with her, persisted during bereavement. From the outside, my friends saw me ricochet through my days, always busy, always energetic. But I was detached from my body, not spending much time aware of my physical state. This prolonged state of agitation likely reflected high levels of stress hormones, which take a toll on the immune system.

In the year or so following my mother's death, I began to experience "twinkling" sensations in my arm. Twinkling was the best description I could come up with for these bizarre electrical sensations that today I would call paresthesia (pronounced paris-thee-zhea). Others might describe these abnormal sensations as feeling like one's skin is being pricked with pins, lasting a few minutes to a few hours. But because they came and went, a few days here and there, I thought little of them, although I did mention them to my doctor at the campus health service of my graduate school. When they roared back, causing me to jump out of bed one night, my doctor pointed out that pains were not normal even if they were intermittent, and they certainly were not "in my head," as I worried they might be. A few rounds of specialists, including sports medicine and rheumatology, turned up nothing. Over a year later, I told my dissertation advisor—a clinical psychologist jointly appointed in departments of psychology, psychiatry, and neurology—about these strange experiences that were leading to so many appointments. He asked me for a fuller description of the pains, and I described the most recent iteration as twinkling sensations in the outline of a hand across my neck and cheek. Given his extensive background in neurology, I think he knew what was causing this, and he suggested I see a neurologist friend of his.

The kind neurologist immediately ordered an MRI, which, in retrospect, I doubt my graduate school health insurance covered. He recognized the sensation as trigeminal neuralgia, a hallmark symptom of multiple sclerosis (MS). Indeed, the MRI showed tiny brain lesions in the places you would expect for someone with MS. "I'm going to give you this diagnosis now and put it in your medical record," the neurologist told me. "We would ordinarily do more tests before giving such a definitive diagnosis, but since you are moving to California soon—well, I can tell you that a young woman such as yourself, with intermittent symptoms, would not receive a diagnosis for far too long." And so I began my clinical internship year at UCLA newly labeled with a chronic autoimmune disorder and no clear prognosis. I doubted I would ever complete my PhD, let alone take the postdoctoral fellowship at UCLA that I hoped to complete in psychoneuroimmunology. Grief over the loss of health is different from loss over the death of a loved one, but the feeling of not knowing how to operate in the world, of wanting one's life put back to the way it should be, makes both of them resound as grief.

Only a few years ago did it dawn on me that the appearance of my MS symptoms might be connected to the death of my mother. And I want to be crystal clear: Bereavement did not cause me to develop MS. My paternal aunt was wheelchair-bound with MS, so I carry a genetic predisposition. I was also raised in the northern latitudes of

Montana, a known risk factor because less sunshine from shorter days provides less vitamin D during child development. And I had mononucleosis in kindergarten and harbor the Epstein-Barr virus, which increases risk for MS. So the death of my mother did not cause my MS, but the stress of grieving was likely a contributing factor to the moment in which the symptoms emerged.

My experience so far with MS pales in comparison with what many people with this disease experience: a mismatch between their physical body and the built world that society has created for the "normal" person, a world where people are not expected to limp or have reduced vision. In fact, without these visible symptoms, I have been able to pass as a person without a chronic illness, and many of my colleagues have not known. I see the struggle with my MS symptoms behind closed doors as similar to what many grieving people have described to me. They stop telling people about how they are really feeling when they are asked (with a sad face), "How are you doing?" They mask the emptiness or loneliness they feel but do not let on, because they feel they should be "over it," and people are uncomfortable around someone who is in pain. Learning how to manage grueling episodic fatigue and other symptoms has given me great empathy for bereaved people who manage the waves of grief that could overwhelm them at any time.

More important to the book I offer you now are the lessons I learned about leading a meaningful life in the face of terrible loss as I teetered through managing a chronic illness. Loss due to chronic illness and loss due to bereavement both involve the unbearable cloak of grieving, weighing on you every day, shaping what you think, what you do, and your beliefs about the future. When I was first diagnosed, I was told that the first five years would tell me a lot about what the course of my MS would look like. *Five years?!*, I thought. It is hard to figure out who you are, and how to plan, when you are uncertain about the future. With a chronic illness, you cannot imagine a future with unspecified disability. With loss, you cannot imagine a future without your beloved, who was always supposed to be there with you.

In this book I share with you my experience of grief after the death of my mother and my diagnosis in retrospect, knowing what I know now after years of hard-won wisdom, learning how to relate to my internal state differently, becoming aware of how I was feeling, and figuring out how to soothe myself. I hope what I have discovered will be helpful to you, who may also have faced the heart-wrenching death of a loved one or another type of loss, and yet wish to make good use of this one life you are given, to share the gift of the love you have known, and to accomplish what you set out to do before your life went terribly off track.

My expertise in explaining the physiology of grief and grieving is based on more than illuminating metaphors or personal experience, however. I have spent the last quarter of a century studying the body's mysteries, in blood samples and heart-rate patterns deconstructed into their component parts. Physiology involves a lot of techniques that take laborious study—the methods of transforming physiological signals into numbers. Learning these techniques includes making sure the data represent the signal you want to detect, and not all the other signals the human body produces simultaneously. If you are measuring the influence of the parasympathetic nervous system on the heart, you do not want to accidentally be measuring the influence of breathing on the heart instead. If you are measuring brain activity, you do not want to accidentally include involuntary head movement as well. In addition, research requires not only interpreting the mounds of data produced but also doing so without making classic errors humans are prone to, by using the scientific method. For example, confusing causation with correlation means that we often assume that when two things occur together, one is the reason for the other, unless we carefully design experiments to tease apart order or timing.

All scientists walk into their labs with preconceived notions about the world around us, and those notions form unintentional biases as we decide what questions to ask, who we include in our studies, how we interpret what we see in the data, and whether we deem our research worthy enough to publish. My research has been no different, sometimes suffering from small sample sizes or nonrepresentative samples or not sufficiently replicated findings. Science bears human fallibility. But the process of thinking through the critiques of my research, conducting hundreds of interviews with grieving people while observing their physiology, and always pursuing the underlying theories that might explain *why* and *how* grieving works, gives me hope that the revelations expressed in this book might serve as a lighthouse for others steering their own investigations, whether for the single case study of yourself or a large laboratory producing grief research for the world. This book is for anyone who seeks to reduce suffering from any kind of loss, and seeks more understanding of us in our world.

I began my own journey by studying psychophysiology in graduate school. I was fascinated by the way our cardiovascular system reacts to the world around us and then recovers, enabling us to experience pain and pleasure, and to regulate our emotions. After psychophysiology, I studied neuroscience and specifically functional neuroimaging to peek inside the tantalizing black box of eighty-six billion neurons. But because the brain is not disconnected from the body, I also became curious about how the immune and central nervous systems work together—and so I began to study psychoneuroimmunology. There is a lot to learn when studying immunology, with a whole new set of techniques and possible errors to be made. The immune system is wondrous, able to learn and remember what germs it has seen before, rivaling the impressiveness of the brain. I told myself that I was done taking on new methods—except that by isolating immune cells from blood samples, cells that harbor all of our DNA, the role of genetics in grieving became tantalizingly available. And so I studied genetics as well, leading to discoveries that I believe are only in their infancy but will help us understand the development of our attachment relationships, our responses to stress, and how we carry the experiences of our ancestors forward. My graduate students have spurred me to conduct clinical interventions, an opportunity to determine whether a specific ingredient in a person's grieving process could be changed and, if so, whether it would lead to an easier road for them. While my pursuit of understanding grief has been dogged, using many different methods and techniques, my goal is to put this knowledge into your hands now, so you can apply the scientific findings to your own life and your own grieving.

We really can die of a broken heart. The undeniable physical response to bereavement emerges in data. In the first three months after the death of his wife, a man's risk for his own mortality is almost twice as high as a similar man who remains married during the same time.[2] The widowhood effect is seen in women as well, at over one and a half times higher risk. We gleaned this knowledge from a huge 2013 epidemiological study in the United States that included 12,316 people. But we have known for a long time that bereavement increases mortality. Colin Murray Parkes, one of the first psychiatrists to specialize in grief, published evidence of the widowhood effect on mortality in the *British Medical Journal* in 1969, and actuarial tables from life insurance companies documented this phenomenon even earlier. From painstaking research, we know the widowhood effect wanes over time. From a study of 1.5 million Finns, we know that between six months and six years following a death, mortality risk for the bereaved drops to only 16 percent higher for men and only 6 percent higher for women than their married counterparts.[3]

The epidemiological evidence for dying of a broken heart has made me wonder about—and study—what physiological changes happen in our body after the death of a loved one that might lead to outcomes as significant as illness or death. The book that follows describes the many organs and systems that may be affected, including the heart, the immune and endocrine systems, the lungs and liver, the brain, and the sympathetic nervous system.

Many people assume the stress of bereavement must be unique, to warrant the study of its physiology. But the way your body responds to stress is quite simple: Your heart rate and blood pressure can go only up or down. Your inflammation level can be higher or lower, and so on. So although the death of a loved one has long been noted as the most stressful life event, the stress response does not have to look *different* from responses to other stressors in order to make bereavement a worthy area of study.

The point is not whether bereavement causes a unique stress response; rather, I am pointing out that beyond the emotional reactions we experience during bereavement, we have a significant physiological stress response. Grieving wails and screams, not just as outward expressions of emotion, but inside our body as well. It is my greatest hope that reading this book will lead you to deeply understand that grief is a physical experience. And that this understanding will also be a call to action. Because if we understand that grief is physiological, there are important implications for how we care for those who are grieving.

Bereavement is a period of increased risk for illness and death, and recognizing this should lead to better medical care. We should treat bereavement as the monumental stress event that it can be. Grief is not a disease, but it has physiological effects, just as pregnancy is not a disease, but it increases the risk of hypertension and gestational diabetes. Consequently, societies develop systems to assess risk and provide care when women are pregnant. Taking the physiology of bereavement seriously, understanding the new medical science this book includes, should lead us to advocate for health care as a necessary response after the death of a loved one.

I have organized this book in two parts, the first part contemplating the physical and even medical impact of grief and grieving, and the second part exploring how to consider grieving as an opportunity for healing.

The chapters unfold as a series of questions:

- Chapter 1 uses research from epidemiology and cardiovascular psychophysiology to answer the basic question, "Can we die of a broken heart?"

- Chapter 2 delves into the complex and highly intelligent immune system, which carefully balances protecting what is *self* with eradicating what is *non-self*. This chapter asks, "Why does the immune system tilt off-balance in the wake of the enormous stress of bereavement?" Here I share my own studies in psychoneuroimmunology, studying inflammation and white blood cells in those who have lost a loved one.

- Chapter 3 wonders, "How do our hormones reflect the stress of cutting away our loved one, and why does bonding mean that this hormonal impact is inevitable following separation?"

- In chapter 4, I turn to clinical psychology and behavioral science to consider, "How do our coping behaviors in turn affect our physical health?"

- For chapter 5, I return to my neuroscience background, but this time to explain how our brain, as an organ like any other, must cope with increased inflammation and remodeling of its cells. But because the brain is also unlike any other organ, we can ask, "How do physical changes in the brain after the death of a loved one alter our emotional and cognitive states?"

- The topic in chapter 6 is the impact of grief and grieving on our peripheral nervous system. I trace the sympathetic nerves in the body and the evolution of viruses and bacteria to ask, "Why are we more prone to illness during times of enormous stress?"

Part Two focuses on the process of healing during grieving:

- In chapter 7, I rely on rehabilitation psychology research to take up the question of motivation, asking, "How can we pursue value-driven goals if we have no energy and fail to see the point?"

- Chapter 8 delves into the science of behavior change and digs for answers to the question, "How can we develop new and healthy habits in the absence of our beloved?"

- In chapter 9, cognitive science and Buddhist practice help answer the question, "How do we raise to consciousness the things we avoid and maintain our attention on overcoming the barriers to meaningful activity in our lives?"

- Chapter 10 explores, "How can we reconnect with others when we feel broken, and how could that help our physical health?"

I hope the work my fellow grief researchers and I have done will continue to elevate our concern for our bereaved friends and families, patients, and clients until we better understand how to help those who have suffered a loss. My aim is to raise awareness that not only is bereavement a stressful life experience but also that grieving is a medical risk factor. We can all learn more skilled responses to suffering during grieving. By sharing with you the physiology of grief, and exploring with you a diverse toolkit of coping strategies, I hope those of you who are grieving may come to discover continuous healing in the service of restoring a meaningful life.

PART ONE

The Physical Nature of Grief

CHAPTER 1

The Heart

J oe Garcia collapsed and died of a heart attack on Tuesday, May 24, 2022. He had just returned home after visiting the memorial of his beloved wife, Irma Garcia, a teacher who was shot and killed two days earlier in the Robb Elementary School massacre in Uvalde, Texas.[1] When we see the photos of the paired black hearses, carrying their two bodies to the cemetery, it resonates with us as an image of deepest grief. Tragically, we can truly die of a broken heart. As a psychologist, I have investigated this most human of mysteries: How does the heartrending loss of a loved one affect us so profoundly as to cause death? What is the evidence that our heart can physically break after a loss?

What we know is that our chances of dying of a broken heart skyrocket as soon as we learn of our loved one's death. The more recent the death, the more impactful for our body. In fact, epidemiologist Elizabeth Mostofsky and cardiologist Geoffrey Tofler studied people who had heart attacks (including nonfatal ones) and discovered that a heart attack was twenty-one times more likely to happen in the first twenty-four hours after the death of their loved one, compared to any other day in their life.[2] Twenty-one times more likely. The risk on that first day is even higher when our loss feels moderately or extremely meaningful, pointing out that the psychological aspect of grief contributes to our medical outcome. On each subsequent day after the loss, our risk of a heart attack decreases.

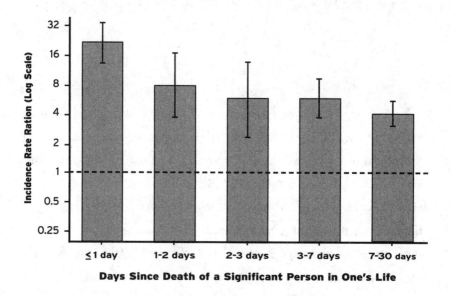

Days Since Death of a Significant Person in One's Life

Although dying of a broken heart has mostly been studied after the death of a spouse, cardiovascular incidents like a heart attack or a stroke happen after the death of a child and other loved ones as well. Perhaps the most famous example is the death of Debbie

Reynolds the day after her daughter, Carrie Fisher, died. And after the death of a parent, our risk of ischemic heart disease increases by 41 percent and our risk of stroke increases by 30 percent.

Regardless of whether the person you lost is your spouse, child, sibling, or best friend, the most important predictor of dying of a broken heart appears to be the strength of your attachment to the person who died. We can see evidence of this from a Swedish study by Dang Wei and Krisztina László, who observed patients after they survived a heart attack following the death of their loved one. These researchers discovered that when heart attack survivors said the impact of their loved one's loss was significant, regardless of their kinship relationship to the deceased, they had a 12 percent increased risk of having a second cardiovascular event (as opposed to when they reported the impact was mild or moderate). The intensity of our grief response comes from who the person was *to us*, rather than just who they were, and it can wring the life out of our heart.

Having a Theory About Why

For me, these incredible data demonstrating that grief increases the risk of one's own mortality raise the question "Why?" Theories as to *why* grief has this impact on the heart seem vitally important to solving this mystery, rather than simply documenting *what happens*. Most research studies consider a whole host of possible contributors, and these explain only a little of the increased risk. For example, in that last study, Wei and colleagues included the influence of age, sex, education, financial stress during childhood, parental smoking during childhood, partner's smoking, survivor's smoking, stroke, diabetes, physical activity, parents' history of heart disease, etc. All these other factors accounted for only a small contribution to the increased risk of a second cardiovascular event after the death of a loved one. This means the broken-heart effect is still there even when we try to explain it away through myriad other physical reasons. The psychological impact of grief shocks our heart; the absence of our one-and-only tears at our very tissue.

Two scientific strategies can help us to discover why grief has this impact. The first strategy is to develop theories that incorporate what we know about both the psychological and the physiological responses in grief, like *attachment theory*. Working from the top down, these theories point us to where to look for changes in biomarkers, and researchers can then design studies to discover whether the data bear out that those theories are true. The second strategy works from the bottom up. We can assess the physiological mechanisms in bereaved people that constitute heart function (like heart rate, blood pressure, inflammation, and arrhythmia). By determining whether bereaved people show changes in these biomarkers, scientists can plot a plausible path from the death event, through these mechanisms, to the broken heart. For example, not all bereaved people experience changes to heart function. So who is at highest risk for the broken-heart phenomenon? If we understand the physiological mechanisms, we may find answers as to whether it is the *severity* of our grief response to a death loss, rather than simply the fact of a death loss, that matters most to our physical health. Of course, the best way forward is to use both top-down theories and the bottom-up accumulation of empirical data together.

Attachment Theory

Throughout this book, I explain several theories as to why we are so seriously affected by the death of a loved one. In my own research, I have relied heavily on attachment theory to understand why a loved one's death might cause us physical harm. Attachment theory was first described in 1982 by British psychiatrist John Bowlby, who made careful observations of infants who were separated from their parents. Relevant to the loss events of his time, Bowlby had studied orphans of World War II and systematically described their protest and then despair when they were orphaned. Bowlby went further and made comparisons to separation reactions in other animal species as well, giving us brilliant new insights into the basic science of why loss has such a huge impact. We cannot study grief and loss without first studying love and bonding, and his studies of separation revealed the invisible tether of attachment between parent and child.

Since I live in the American Southwest, when I think of attachment, I always picture a string of little quail following in a line behind their bigger mommy or daddy quail. Each baby quail has a tiny little feather at the crown of its head, bobbing up and down, a miniature version of their parents' same bobbing feather. One year, quail nested in a huge flowerpot on my back patio. At some point in the spring, I heard a racket outside and looked through the sliding-glass door to see one of the adult quail standing on the ground, loudly encouraging the little ones to make their first leap out of the pot. This took quite a bit of time and encouragement, especially to get the last chick to venture out. Protesting separation, and making the effort to stay together, is part and parcel of our attachment bond. That bond is the invisible tether, sometimes stretching across miles or across days and weeks. It is the deeply held belief "I must wait for them to return. They will always come back for me" or "I must find them. I know they are out there waiting for me." Our implicit beliefs arising from attachment motivate incredible patience or effort. This motivation gets its strength from powerful neurochemicals in the brain. Once we have formed a bond with our loved one, our brain understands that this one-and-only is different from any other of our species and uses dopamine, opioids, oxytocin, and cortisol to motivate us to seek them out.

Attachment theory predicts two responses to the absence of our loved one, demonstrated in numerous research studies: protest and despair. Seeking out our one-and-only, along with protesting being separated from them, clearly makes sense when our loved one is alive. If loudly protesting their absence does not work, if it does not bring them back, then despair settles in. But even our despair has its function—this is our brain recognizing that the one-and-only will never return, and despair is the manifestation of this deep knowledge of our new reality. I think of protest as *"Oh no, they're gone?!?"* and despair as *"Oh no . . . they're gone . . ."* Both reflect awareness of the new reality, as painful as it is.

But this is not the end of the story. This natural learning process eventually allows us to return to living in the present moment, to transform our relationship with the deceased through continuing bonds, and potentially to strengthen our attachment relationships with living loved ones or even form new bonds. For most of us, our physical body can withstand the stressful process of grieving, despite how painful and disruptive it is. Our attachment neurobiology is designed to resiliently allow us, with support, with time and experience, to continue to pursue a meaningful life, having gained the bittersweet awareness that love and grief are all a part of it.

The Heart Is a Remarkable Organ

Although most of us think of our heart in terms of plumbing (blood pressure through pipes, corrosion in arteries), the most fascinating aspect of our heart is electrical. Pacemaker cells autonomously beat to their own drum of approximately 220 beats per minute in each and every heart, for no apparent reason other than the miracle of life. Two primary electrical systems wire—or innervate—the heart: our sympathetic nervous system and our parasympathetic nervous system, together called the autonomic nervous system. The parasympathetic nervous system uses the vagus nerve, a word that comes from the same root as *vagabond,* because the vagus nerve "wanders" throughout our body, regulating many of our organs in addition to the heart. This same set of nerve fibers controls our larynx, trachea, stomach, pancreas, kidneys, and intestines.

Our parasympathetic nervous system has captured my attention since the early days of my graduate training, when I studied how this system might react during bereavement. If our sympathetic system can be thought of as the gas pedal, the vagus nerve of our parasympathetic system can be thought of as the brake pedal. The reason our heart ticks along at a resting heart rate of less than 220 beats per minute is because our vagal brake is constantly being applied (at least when we are sitting still). The genius of this system is that when our vagus nerve lets off the brake a tiny bit, the heart immediately accelerates. For our sympathetic system to get the heart going faster, a great input of energy is required to release the hormone adrenaline. But for our vagus nerve, it takes much less energy to simply let off the brake to increase the heart rate. So our parasympathetic system can react much more quickly and with less effort. Notably, the sympathetic and parasympathetic are not always reciprocal. They can both be activated, analogous to a drag racer at the starting line, pressing both pedals so the car is still until the flag goes down. These two systems work together to match our heart rate to what our body needs, based on what is happening in our environment.

Why does a psychologist study the unique influences of sympathetic and parasympathetic influences on the heart? Because emotion, concentration, and recognition of stressful events all affect our heart rate. Psychophysiology has been a part of psychology dating back to the earliest laboratory of German physiologist Wilhelm Wundt, when "psychology" was a new discipline in 1879. Assessment of heart rate has come a long way since then, utilizing elaborate mathematical models to evaluate the way one heartbeat relates to the next one in the sequence. Picture the readout of your heart rate on graph paper, the sharp peaks etched on an electrocardiograph. Now imagine how far apart those peaks are—for a fast heart rate, the peaks are close together, and when the rate is slow, the peaks are farther apart. But the distance between peaks is almost never constant.

In fact, our heart rate continuously accelerates and decelerates a tiny bit or exhibits *heart rate variability*. This acceleration and deceleration follows our breath. When we inhale, our heart rate goes up. When we exhale, our heart rate goes down. If you go running and get your heart pumping, you can feel this yourself by taking your pulse when you stop. During your inhale, your heart rate is fast from your exertion. During your exhale, you will feel your pulse slow. Understanding these influences can help us to solve the mystery of why a stressful event might overwhelm our heart following the death of a loved one.

But first, the knowledge of cardiovascular functioning can help us to understand how bonds with our loved ones are created and maintained. Social engagement, and the cardiac control it requires, begins when we are infants. Much of this happens during nursing, which necessitates a unique behavioral state, as infants must focus and remain nearly immobile. I have recently had the joy of taking care of an infant, and I can tell you that sucking from a bottle requires a surprising amount of effort and concentration!

Social engagement requires us to make eye contact, modulate our tone of voice, and form facial expressions, all of which become associated with the calm and safe experience of nursing. Our ability to know when to trust other human beings begins here, based on what we see and hear from our caregivers. Our capacity for social bonding starts with cooing, gazing, smiling, and sucking, all regulated by our autonomic system. Through these expressions, we communicate whether we want our loved one to come closer or to give us some space, and we learn the signals from our loved one for the same requests. Reading social cues, expressing our emotions, getting riled up, and then calming down again—each of these are aided by our autonomic nervous system. Together with sympathetic nerves, the parasympathetic nervous system enables calming (rest and digestion) and communicating (vocalizations, facial expressions, and head rotation). This well-regulated social engagement enables the creation of our attachment bonds. Our psychology and physiology are inextricably linked.

Better Vagal Tone Facilitates Bereavement Intervention

How does our vagal brake relate to bereavement stress? In my dissertation study, I was curious as to whether individual differences in vagal control would enable better cardiac regulation during waves of grief. We know from research with nonbereaved people that more effective vagal control predicts fewer heart attacks. So I asked bereaved participants to come into the lab for my study, and I measured their heart rate variability at rest. Then I asked them to write about what happened when their loved one died. I asked them to come back the next week and write a letter to their loved one, saying goodbye. The third week, I asked them to write about how the death of their loved one had affected their life.

I discovered that those with greater vagal cardiac control showed decreased grief across the three weeks. They benefited most from processing their grief experience through writing, compared to those bereaved people with less vagal control.[3] I also studied a group of bereaved people who came in for the same three weeks and just wrote unemotionally about the facts of what had happened during their day. These latter participants served as a comparison group, so I could distinguish the writing intervention effect from any change that occurs naturally during grieving. Unlike the intervention group, improvement in grief in the comparison group was not predicted by whether or not participants had high vagal control.

I interpret these results to mean that two different elements are important in grieving: emotional processing *and* concurrent physiological regulation. Expressing our feelings and thoughts about grief enables us to grapple with what has happened and what it means for us. If our cardiac regulation enables us to react and recover from these waves of grief, then we benefit from that emotional expression in the form of less grief over time. This study demonstrated the interrelated links between our physiology and our psychology. The heart can be thought of as an accurate barometer of the nature and intensity of our thoughts, feelings, and motivations. From the readout of the heart, we can see the automatic, unconscious dance between our body and the loss that shattered our world. Seeing the effect of grief in the peaks and valleys of an electrocardiogram reminds us that the impact of loss, although it seems mysterious, is real.

A Broken Heart vs. Takotsubo Cardiomyopathy

The increased rate of mortality among bereaved people we have been discussing, or "dying of a broken heart," is different from Takotsubo cardiomyopathy. Takotsubo cardiomyopathy, first reported in 1990 by Hikaru Sato and colleagues while treating Japanese earthquake survivors, is defined by a specific anatomical "ballooning" of the heart. When people with Takotsubo cardiomyopathy show up in the emergency room, it looks for all the world like they are having a heart attack. But they are not having a heart attack, although they still need medical attention.[4] So the broken-heart phenomenon is broader—it *includes* Takotsubo cardiomyopathy but also heart attacks (including fatal ones), strokes, and other medical events. Dying of a broken heart means that the mortality rate among bereaved people is higher than nonbereaved people, regardless of the cause.

Takotsubo cardiomyopathy is nonetheless fascinating. It usually follows an extremely stressful event in the person's day. This stressful event could be the death of a loved one but can also be a significant argument, learning of the illness of a close person, significant stress at work, attending a funeral, or an earthquake. Less widely known, and much rarer, Takotsubo cardiomyopathy can also occur after an intensely positive event, like a family wedding, a milestone birthday party, the triumph of a favorite sports team, or winning the lottery.[5] The temporary ballooning of the heart in Takotsubo cardiomyopathy following an emotional event likely results from the "stunning" of the heart by adrenaline. While Takotsubo cardiomyopathy happens on the day of an emotional event, dying of a broken heart can occur at any time after the loss, from the same day to weeks, months, or even years later.

An Emotional Stress Test for the Heart

A unique study from Australia by Thomas Buckley and Geoffrey Tofler investigated the mechanisms through which bereaved people might die of a broken heart.[6] Buckley, a critical care nurse, contacted bereaved spouses or parents of deceased patients from hospital emergency departments within seventy-two hours following the death of their loved one. Participants in the study had cardiovascular assessments within two weeks of the death and again at six months. Buckley also enrolled, as a comparison group, nonbereaved family members of patients from other hospital departments (like orthopedic or endoscopy clinics) and assessed them at two weeks and six months as well.

From this careful study, we know that acutely bereaved people have a higher twenty-four-hour heart rate and higher blood pressure, and lower heart rate variability than their nonbereaved counterparts. These are known cardiovascular risk factors, suggesting they are mechanisms that could contribute to the increased cardiovascular risk in bereavement. In addition, the bereaved showed risk factors for heart attack and stroke, prothrombotic changes seen in their blood samples. Here is the most valuable part of Buckley's longitudinal study: These cardiovascular measures had returned to normal after six months, so that heart function of bereaved and nonbereaved groups was indistinguishable at that point.

Roman Palitsky (a graduate student of mine) and I asked: If grief is causing our heart to malfunction, when is this happening? Under what circumstances does the heart attack occur? Palitsky focused blood pressure as a mechanism that changes during bereavement and may contribute to the broken-heart phenomenon. Our idea was that the heart is most likely to have trouble during a wave of grief. Much like a treadmill stress test in a cardiologist's office, we wanted to see how the heart responded during a wave of grief, a kind of emotional stress test.

In order to create a wave of grief, Palitsky and our team used a grief recall task to evoke feelings of grief while bereaved research participants were hooked up to an electrocardiograph and a blood pressure monitor. The grief recall task comes from a long line of research that cardiologists developed to assess the effect of other emotions on heart function. Originally, the anger recall task helped researchers to discover the role of anger in cardiovascular response and recovery, by asking participants about a time when they felt enraged and then maintaining their attention on this memory for ten minutes. We adapted this to grief by asking participants about a time when they felt very alone since the death of their loved one. Blood pressure was evaluated before and after this ten-minute wave of grief.

Participants' systolic blood pressure, or the pressure that the heart exerts on the arteries while beating, increased from before the task by approximately twenty millimeters of mercury (the unit used to measure blood pressure) to after the grief recall task.[7] Thus, the fact that waves of grief affect our blood pressure, while not surprising, was clear. But it might surprise you to know that blood pressure increased on average as much as it does with moderate exercise, even though the participants were simply sitting there, talking with us.

Even more important, the participants who walked into the lab with the highest level of grief symptoms were the ones who experienced the greatest increase in their blood pressure during the wave of grief. For those with the lowest level of grief prior to the task, systolic blood pressure increased about ten millimeters of mercury, but those with the highest level of grief showed an increase of twenty-four millimeters of mercury. These data mean that it is not just the death of a loved one but also our emotional response to that loss that affects our heart. Assessing grief severity also gives clinicians a way to identify bereaved people who might be at highest risk for heart problems, by asking about a specific set of grief symptoms occurring over the past two weeks. Most people experience normal waves of grief during acute bereavement, and thus, we need more research to determine the benchmark level of grief severity that predicts increased cardiovascular risk. But regardless, finding a way to calm ourselves following waves of grief may potentially be helpful.

Please do not misunderstand me: I think waves of grief are natural and healthy. But we can benefit from both the rise and the recovery of our emotional surges, like the wave that crashes on the beach and then ebbs away, back into the ocean.

Absence Makes the Heart Respond

I had never been alone overnight until I was eighteen, the summer after I graduated high school and before I went off to college. I took a job as a live-in nanny with a family about three hours from the town where I grew up. One weekend, the family went away on a mini vacation, leaving me on my own overnight for the first time ever. As that first evening drew closer, I was overcome with an inexplicable experience. I could not put my finger on why I was having such intense feelings or even what the feelings were. I vividly remember trying to figure out what was happening to me, and the closest thing I could compare it to was the feeling I had before I went onstage for a music performance. It was excruciating. And mystifying, as the feeling seemed to have no cause, and it went on and on. I tried cleaning the house, vacuuming vigorously in response to my seemingly endless restless energy, as though I had drunk a whole pot of coffee. I finally called someone I had met in the neighborhood, and she let me stay overnight at her place. What I recognize now as separation anxiety would plague me for the next few decades, an unnamed terror I dreaded.

What I would not learn for decades was that this kind of anxiety is a common response to being separated. All human beings have a neurobiological attachment system, even though not everyone reacts the same way to being alone. Being apart from a loved one provokes a response in us, and those feelings are normal and adaptive. This same reaction to the absence of our one-and-only reveals our bond with them. Attachment ultimately brings bonded people back together after being away, enabling them to explore the world, seek out food or discover new vistas, and return to their loved one. Attachment is vital to our survival, and that panicky response to being separated motivates us to reunite with them. Returning keeps us safe, offering us protection and nurturance.

Perhaps it is no wonder that neuroscientist Jaak Panksepp called the neurocircuitry underlying this separation response the PANIC/ GRIEF system. You have probably also had this response: Have you ever looked down and realized that your child is not at your side? The adrenaline that suddenly courses through your veins, and the resulting frenetic searching, is the recognition and response by your neurobiological attachment system to the fact that a vital part of you is missing. You and that other person have formed an "us," which is now the baseline for how you function in the world. Because that bond has been formed, the absence of the "you" from the "me" means that the "us" is not whole, and engenders a response in proportion to how central this person is to your functioning. Part of this response is an increase in heart rate, enabling that protest response, calling out and searching for them. The increased heart rate is a useful response to separation, giving us the boost in energy needed to search for them. But in situations where separation is the permanent new reality, like in the uncommon situation of the death of a loved one, that increase in heart rate still plagues us. Our brain has an imperative: keeping our loved ones close. Our new reality of permanent loss will only be learned slowly, as our grieving over time enables us to manage our waves of panic and grief, develop new attachments and/or strengthen existing ones, and find meaning again in the new world we now inhabit.

Unfortunately, because most of us do not know how the powerful attachment system works in human beings, we find our physiological responses (increased heart rate) and our behavior (restless energy) to be mysterious when we are separated from loved ones. I was at a complete loss to name what I was feeling that summer as night grew closer and no one was in the house with me. The neighbor who let me stay over assumed I was afraid of having the house broken into. I knew I was not afraid of something dangerous happening, but it never occurred to me that my panic had been caused by the absence of someone close. Moving away from home was something I wanted desperately to do—there was so much out in the world to explore, and I was driven by my developing brain and body to want those pioneering experiences. But separation from my mother was not yet something I had learned to tolerate. Through my childhood and adolescence, she also could not tolerate separation from me, and so I had few experiences to call on and few developed skills for coping with feeling alone.

Our Loved One as Our External Pacemaker

To understand the mystery of why we physically suffer upon separation from a loved one, of how our heart could break when they die, we must first understand the role they play in our physiology when they are alive. Co-regulation defines a relationship. When "you" and "me" become bonded, our "us" is now a functional system, and our physiological systems (cardiovascular, endocrine) are entwined, even though they operate in our separate bodies.[8] Like the invisible force of gravity, attachment is a property of two people held in each other's orbit. Viewing an attachment relationship as a dynamic system, psychiatrist Myron Hofer wrote a groundbreaking paper in grief research in 1984, called "Relationships as Regulators: A Psychobiologic Perspective on Bereavement." He pointed out that this regulation is not apparent while the two members of a dyad are going about their daily lives. Only by removing one member of the dyad is the regulatory role revealed, because the absence of one affects the functioning of the other. In fact, the relationship serves a regulatory role for several different physiological systems.

For example, Hofer discovered that in two-week-old rat pups, physiological regulation is delegated in part to the mother. When the mother was removed, if researchers continued to provide warmth to the pups, the slow decline in activity that is usually seen after separation was prevented. In addition to providing thermoregulatory input, by providing enough milk for appropriate weight gain, normal heart rate was maintained in the pups. Providing this milk on the established nursing schedule of the mom during a twenty-four-hour cycle also regulated the rapid-eye-movement (REM) sleep of the pups. This meant that several regulatory relationships were revealed: individual aspects of Mom's input regulated different physiological systems of the pups.[9]

This last example of diurnal-rhythm regulation points out an important role we fill in the lives of our loved ones—our systems are in sync, entrained, regulating one another. Our biological systems run on a diurnal pattern, our hormones rising and falling throughout twenty-four hours in a predictable way, entrained with light and dark because we have evolved to live on planet Earth. Sleepiness and hunger also have this diurnal rhythm. This is all orchestrated with complex oscillators in the hypothalamus of our brain, pacemakers that are continuously increasing and decreasing hormone production across the day and night. But these pacemakers require daily synchronization. Left on their own, without any environmental inputs, they would run on something closer to a twenty-five-hour cycle. The environmental inputs our brain uses for the small adjustments that keep us on a twenty-four-hour cycle are called zeitgebers, which means "time givers" in German. Many zeitgebers have been discovered, including light/dark and temperature. Most fascinating, many zeitgebers are social inputs like our loved ones, and these entrain us both to each other and to our natural world.

When a loved one dies, their inputs, which have invisibly regulated our body, are torn away all at once. We are dysregulated, without their smell and touch and timing. This dysregulation shows up as mental fogginess, or may be experienced as fatigue or restlessness, or as a lack of appetite or an inability to sleep. The regulatory role of our one-and-only in our life has been invisible, but their absence reveals that role as clear as day. We may feel we are losing our mind. We discover that completely independent self-regulation of our body does not exist, even in adulthood. Relationships continue to play an important role in the everyday regulation of our internal biological systems throughout our lifespan.

The great wonder of our human body, however, is that it continues to adjust, to heal, to learn. When it becomes clearer and clearer to us that the painful reality of loss is true, that our loved one is not returning, slowly our physiology makes use of the world around us. As we make greater use of the world around us, we become entrained to the new rhythms of life. Because life continues on, and begs us to come with it, it reaches into the very core of our body and helps us to find new equilibrium eventually.

Acknowledging Grief as a Cardiovascular Risk Factor

Although the death of a loved one is a rare event in the life of an individual, it is a nearly universal experience across the population, making the effect of bereavement on medical outcomes a significant public health concern. The association between bereavement and risk of mortality is as strong as other well-established associations, like the association between smoking and mortality. Despite widespread evidence of the increased risk of cardiac events shortly after the death of a loved one, there has been no large-scale attempt to study preventive treatment for the broken-heart phenomenon.

A German medical student, Sebastian Karl, came to my laboratory to tackle this issue of what might be done to prevent cardiac events stemming from grief. He pointed out that acute bereavement as a cardiovascular risk factor is both temporary and predictable. When a person dies in an intensive care unit, a palliative care unit, an emergency department, or a nursing home, we know who the primary caregiver is, standing by the bedside. Our medical institutions usually do not have a holistic view of care, but it is possible to consider that while the patient has been our primary concern up until the moment of their death, the person bonded with the deceased becomes our patient at that moment.

Karl and I came up with the idea that a daily low-dose aspirin during acute bereavement—the brief period with the highest risk—might be a suitable prevention strategy. Aspirin, with its anti-inflammatory properties, targets the main cardiovascular risk factors elevated during bereavement. I liked the idea of using aspirin because it is inexpensive, is widely available, and does not require a prescription, plus the public already associate baby aspirin with prevention of a heart attack. It also has a low incidence of contraindications, other than for people with a history of ulcers or an allergy to aspirin. Karl and I decided that despite a lack of funding, we would try a study as a proof of concept.

For this study, we recruited ten participants bereaved in the past thirty days and twelve nonbereaved individuals, providing aspirin to half of each group and a placebo to the other half. This was not a sample size to prove anything about the efficacy of aspirin for prevention in bereavement; it was an opportunity to demonstrate that acutely bereaved people might find the idea acceptable, and be willing to take a daily pill for a week. During a laboratory visit after participants had taken aspirin or the placebo for a week, we used the same grief recall task described earlier to evoke a wave of grief. Participants who had received aspirin recovered faster after the elicited wave of grief than participants who had received the placebo. For example, heart rate decreased more in the aspirin group, although the small sample size makes it impossible to say whether this would be true in general or was specific to our participants.[10] A larger study of eighty-five participants using both aspirin and a beta-blocker for six weeks of acute bereavement was conducted by Buckley and Tofler in Australia. They also found greater reduction of cardiovascular risk indicators in the medication group, including a lower twenty-four-hour heart rate and lower systolic blood pressure.[11]

As a primary method of preventing heart attacks in the general population, daily aspirin is no longer used. The benefit of preventing cardiovascular events must outweigh the risk of bleeding caused by increased aspirin intake, and that is not clearly the case with long-term use. But acute bereavement increases the cardiovascular risk briefly, and this might shift the equilibrium toward benefit if aspirin is used only for a limited time, like a few weeks. I need to be very clear, however: randomized clinical trials evaluating the effectiveness of short-term aspirin use in acute bereavement have yet to be conducted.

A Call to Action

Attachment theory suggests we give our heart away to the one we love. Once we form a bond with someone, we also become an external pacemaker for their heart, and they become one for ours. This physiological co-regulation has all sorts of benefits when we are together, but upon the death of one member of the couple, the survivor must learn to regulate their cardiovascular system without this relationship. Tragically, losing a part of the "us" may prove too hard on some hearts, and the loss can lead to rapid heart rate, arrhythmia, Takotsubo cardiomyopathy, and, for some, dying of a broken heart. Longer term, bereavement can also affect our immune system, as we will see in the next chapter. I hope awareness of the fact that acute grief is a risk factor for heart failure and other illnesses leads to a campaign of education for our emergency departments, palliative care teams, and funeral directors, to provide holistic care for the bereaved.

CHAPTER 2

The Immune System

My friend Lizzie Pickering wrote *When Grief Equals Love: Long-Term Perspectives on Living with Loss* after the death of her six-year-old son, Harry, from a severe form of muscular dystrophy. Lizzie remembers that, in the first couple of years of grieving, she always seemed to be suffering from a chest cold. Her doctor would prescribe antibiotics, and when she did not improve, she would be diagnosed with pneumonia and given several rounds of stronger antibiotics. She had bad episodes of choking. She just could not breathe.

Lizzie recalls that breathing had begun as an issue as Harry's illness progressed. She was aware of every breath her son took, completely focused on his symptoms and responding to every need his body could not manage. This caring devotion brought her into sync with him, so that as his breaths grew slower, hers did as well.

She became vividly aware of holding her breath. This synchronization meant she was not entirely surprised when she developed pneumonia, because Harry had pneumonia when he died.

In addition, her coping response to his tragic loss was to throw herself into working for the children's hospice where Harry died. It was meaningful work, and she recalls that the only way she felt she could survive was to power forward. Deep in the throes of grief, she worked incredibly long hours, even as she was constantly developing and recovering from bronchitis or pneumonia. Looking back, she recognizes that her body was screaming at her, *Give in and allow yourself to grieve!* But she couldn't.

More than twenty years later, Lizzie has put into effect preventive measures. She sought information about the physical costs of grief and knew she had to make changes in her life. She practices yoga every morning before she starts her day. Although the physical exercise is important, she says the most important benefit comes from grounding herself in the present, tuning into how she is feeling, listening to what her body can do today. Looking back, she wonders how she survived. But grieving has meant she discovered how her body responds to stress, sending her into panic and flight, and how awareness and practice can help her breathe through the waves of grief instead.

Increased All-Cause Mortality

In the previous chapter, we talked about the increased rate of heart attacks and strokes after the death of a loved one. Although cardiovascular causes make up a large portion of the increase in mortality risk, increased rates of cancer (lung, stomach, breast, etc.) and respiratory diseases are also revealed in large studies of the bereaved population. Studies confirm that, like for Lizzie, bereaved people are more susceptible to pneumonia and influenza. In my early research as a postdoctoral fellow, I considered the bewildering array of all-cause mortality after the death of a loved one. It occurred to me that one possible change affecting all of these different illnesses could be in the immune system. I set out to do my own studies, and I found increased inflammation in bereaved research participants as compared to nonbereaved participants.[1] Chris Fagundes at Rice University and others have now done studies with larger sample sizes and find this same increase.[2] Because inflammation can affect any organ and system in our body, greater inflammation could contribute to the many types of increased diseases and disorders we see in the bereaved.

Not everyone develops an illness after the death of a loved one, however. One possibility is that some people have preexisting vulnerabilities, exacerbated by the stress of bereavement. For example, Lizzie says lung infections are a weakness in her father's side of the family, although she had never suffered in that way before Harry's death. The stress of bereavement, with accompanying increased inflammation, might reveal or accelerate existing vulnerability. Vulnerabilities might differ between people, so that one person might struggle to fight off the flu and another person might have a recurrence of cancer. Perhaps my own diagnosis of MS is an example of this vulnerability-stress interaction. I might have developed MS even without the impact of bereavement, since I had a family history of MS. Of course, my individual health history does not constitute a scientific study—I will never know what a parallel life in which my mother did not die would look like for me. But research has clearly indicated that our immune system can be affected during bereavement.

Inflammation

One of the most visible signs of an immune response is inflammation. Even the ancients knew the importance of inflammation and named it in their writings. That universal inflammation response we see on our skin when our protective barrier is breached by a cut or scrape has been noted by humans since we first evolved. Consequently, the hallmark signs of inflammation have ancient names: rubor (redness), tumor (swelling), calor (heat), and dolor (pain). These changes at the site of the breach are brought about by the influx of fluids (blood and lymph) and immune cells (white blood cells) rushing in to fight off any potential infection.

Bacteria, viruses, and parasites are recognized immediately as *non-self* by our skin's resident immune cells, hanging around waiting for action. When they recognize a particle that is not a part of the *self*, they do their best to get rid of it, often by simply gobbling it up and destroying it inside their own cell membrane. This action of recognizing and ingesting that thing-that-does-not-belong also causes immune cells to send messages to the rest of the immune system for help. Those messages are dispatched in the form of proteins called cytokines. This is why cytokines are often called inflammatory markers, because they are the indicators of inflammation.

These resident immune cells in our skin have another role as well. Once they have gobbled up the thing-that-does-not-belong and torn it to bits with their inner cell machinery, they display a trophy of their catch on the outside of the cell. This little bit of the bacteria or virus (called an antigen) cannot hurt us, but it allows the immune cells arriving in response to the cytokine alarm messages to know immediately what they are dealing with. *Ah, this is what we are looking for! This is what has breached the barrier, and we must now search for it and destroy it.* But a few of our resident cells displaying their trophy also make an important journey, heading to the site where they will find the highest concentration of immune cells—what I like to call Lymph Node Discotheque.

Some of our immune cells, like T cells and B cells, do not reside in the skin but circulate throughout the body. T and B cells circulate through our blood and lymph systems, systems that weave through every tissue and organ, providing surveillance for non-self molecules anywhere and everywhere. All of our immune cells detect self and non-self. Immune cells are born in bone marrow and educated there about what constitutes self and non-self. During this developmental phase, they are introduced to every molecule that is native to the body, or self. If a cell sees anything later in its life that is not one of these self molecules, then it knows to attack it. But our T and B cells are special—I like to say that they also earn a college degree. They go to Thymus University and have specialized training in what to do in response to non-self molecules.

Although immune cells destined to take up residence in our skin (and other tissues) will react to a wide range of non-self molecules, each of our T and B cells will react only to one special molecule in the whole universe, it's one-and-only molecule. This incredible complexity, this diversity across the host of immune cells we each produce, happens through an infinite rearrangement of bits of our genetic code. This diversity means that the range of T and B cells in our body can react to any molecule we will ever encounter in our lives. I have taught a psychoneuroimmunology course for many years, and I still find this fact astounding.

I like to tell my students that T and B cells hang out at the Lymph Node Discotheque because for them to have the best chance of finding their one-and-only molecule, a good strategy is to hang out at a popular destination. The immune cell residing in the skin gobbles up the non-self molecule, then displays the trophy antigen molecule on its cell surface before arriving at the lymph node. The special dance between the antigen-presenting immune cell and the one-and-only T or B cell activates the T or B cell to set off an incredible sequence of events. It starts dividing rapidly, creating more cells exactly like itself to react to the intruder. In the case of T cells, the additional cells specific to this bacteria or virus head to the source and help to destroy it. In the case of B cells, the additional cells pump out antibodies that can also go to the source. Ever have a swollen lymph node? During an infection, lots of immune cells are there, and T and B cells are rapidly dividing, making it bulge. When an infection is in the lungs instead of in a cut on the skin, like for Lizzie when she had pneumonia, extra fluid full of immune cells can cause choking and make it harder for a person to breathe deeply.

When the Immune System Goes Awry

There is a downside to the incredibly elegant complexity of our immune system. Our body produces an infinite range of immune cells, able to respond to any molecule in the universe. However, this means that some of our immune cells, when they are born in bone marrow, will react to molecules that are self, like myelin or connective tissue. This means a critical step in the process of our immune cell development is the discovery and removal of any immune cell that will react to a molecule native to our own body. This expulsion must happen prior to the cell's graduation into circulation, while it is still in Bone Marrow Elementary (for all immune cells) or Thymus University (for T and B cells). But any detection system has its errors, and occasionally a T or B cell does slip through and react to our own body's molecules. These errant cells are the potential causes of autoimmunity.

The inflammation that I was describing after a cut on our skin can happen in any tissue of our body, although it is harder to see than on our skin's surface. For example, inflammation happens in our lungs where bacteria have infected lung cells, like in Lizzie's case. Or inflammation in our heart can come from responding to dead cells after organ damage (because those cells are now recognized as non-self) in order to initiate cleanup and removal. Inflammation is a part of most major illnesses—heart disease, cancer, and autoimmune disorders such as rheumatoid arthritis, lupus, and multiple sclerosis.

In multiple sclerosis (MS), T cells mistake our own myelin for non-self molecules. Myelin is the fatty substance surrounding and insulating nerve fibers. This damage to myelin, and the failure to replace it, means nerves conduct electrical signals more slowly in the brain and spinal cord. Without myelin to protect the nerve fibers, the nerve fibers can also be damaged. Damage to either myelin or nerves interferes with the transmission of nerve signals between the brain and other parts of the body. This is why the symptoms of MS are so variable—nerves control almost all the functions of the body, including vision, walking, bladder and bowel control, and cognition. Erroneous nerve signals can result in pain (which was my first symptom), spasticity, or dizziness. Perhaps the most common symptom, however, is overwhelming fatigue, which can limit the activities of people with MS who would otherwise function well.

As I mentioned in the introduction, I experienced the first strange sensations that signified MS about a year after the death of my mother. Although some physiological reactions in grief are acute, like heart attack and increased blood pressure, other physiological grief reactions are more prolonged. For example, inflammation increases in the aftermath of bereavement, and can remain high for a long time. But because inflammation affects many organ systems, the impact of this heightened inflammation can lead to persistent changes and may have affected me long after my mother's death.

The History of Immune and Bereavement Studies

I am not the first researcher to ask whether bereavement causes changes in the immune system. In 1977, Australian psychiatrist

Roger Bartrop published in *The Lancet* the first study on the immune system's response to the death of a loved one. Over the next forty years, about three dozen additional studies were completed, investigating many different aspects of the immune response. Some studies looked at whether the number of immune cells declined; others looked at whether immune cells did not function as well as in nonbereaved people. Still others looked at the ability of a bereaved person's immune cells to react to bacteria in a glass dish (in vitro), and others looked at whether bereaved people had increases in inflammation. In short, this area of grief research had many stops and starts. No unified theory or even systematic process was used to tackle the unwieldy question of whether the complex human immune system is affected in bereavement. Nonetheless, these studies built a foundation. In 2019, psychologist Lindsey Knowles and I reviewed all of these papers, and showed that although some of the early studies had small samples of people and used older methods of evaluating the immune system, most of the studies showed a connection between poorer immune function and bereavement.

Such studies reveal that the death of a loved one may change the constellation of our immune cells in circulation, decreasing the proportion of some types of immune cells. Additional evidence suggests that immune cells become less effective at mounting a defense against viruses and bacteria. For example, older widows and widowers benefit less from an influenza vaccine compared to nonbereaved older adults, resulting in less immunity to the flu.[3] In fact, after the death of a loved one, flu, pneumonia, and sepsis infections are all more likely.

When Knowles and I read the span of the scientific literature on bereavement and the immune system, even though the studies were scattered across several decades, we were struck by a theme that is not revealed in studies of the acute broken-heart phenomenon. Bereavement may be a risk factor for increased risk of disease and death, but a specific group within the bereaved accounts for the majority of the risk—those who show the most severe grief response. Bereavement is a category—you are either bereaved or not bereaved within a defined period of time. Like ticking the box *widowed*, everyone who is bereaved is lumped together in the epidemiological data. But rather than asking simply if you are bereaved, we can ask about the severity of your grief reaction. By giving a bereaved person a self-report questionnaire (sometimes called a scale) or by interviewing them with a set of standardized questions, we can determine the *intensity* and *frequency* of the person's waves of grief and how much their grief interferes with their day-to-day ability to function. The severity of their reaction to the loss tells us something different than a yes-or-no question of whether they have had a loss. Our review of immune studies suggested that severe grief or depressive symptoms during bereavement were likely to impair immune function more than typical resilient responses to the death of a loved one.

It is important to understand the impact of both bereavement more generally and grief severity specifically. Each may predict different medical outcomes. I suspect we may find bereavement predicts acute medical events, but grief severity predicts longer-term health problems. If those with the more severe grief reactions to loss are at higher risk for poor medical outcomes, our healthcare system should screen for this, and follow them or provide preventive care.

The Theory of Sickness Behavior

We have been discussing the way that the stress of bereavement might contribute to inflammation or immune dysfunction. On the flip side, increased inflammation might make our grief feel worse. A good theory would help us to predict what might happen under different bereavement conditions, and give us insight into how to intervene with people struggling with both the devastating emotions of grief and an increased vulnerability to illness. Most fascinating, a theory might reveal that some of the psychological impact after the death of a loved one could result from our immune system directly motivating our behavior.

In the first chapter we discussed attachment theory. A second theory that has helped me understand how grief works emerged in the 1990s from the new field of psychoneuroimmunology (PNI). The *theory of sickness behavior* sprang from the discovery that when inflammation increases in the body, it changes our behavior, our goals, our whole motivation. Remember the last time you had the flu? You felt "awful," but more specifically, you probably had very little interest in doing anything social, in going to work or taking an exam, wanting to just hunker down in bed. PNI researchers realized that when people have high levels of inflammation due to illness, they often feel blue, have difficulty sleeping and eating, and withdraw from other people. Although it is typically short-lived, this looks a lot like depression. In a different example, when patients start a type of chemotherapy that increases inflammation, we see a dramatic increase in symptoms of depression. This points to inflammation as the cause of the increase in depressive symptoms. In fact, for some kinds of chemotherapy, the standard of care is now to provide antidepressants prior to initiating treatment. Even in animals, if we administer inflammatory proteins directly into their system, we see their exploration and social interaction decrease. How we feel emotionally is often motivated not only by what we think but also by how we feel physically (even outside of awareness). High inflammation influences our behavior in ways we can clearly demonstrate and measure in scientific studies. Given we know that inflammation increases during bereavement, perhaps inflammation can help to explain some of the changes in how we feel, think, and behave while we are grieving.

Our Body's Response to the Injury of Loss

Attachment theory identifies our protest over the loss of a beloved person as coinciding with increased heart rate, blood pressure, and cortisol levels—the survival mode that often initially takes over. This protest response includes preoccupation with the lost person. We feel anxiety and may search for them, trying to re-establish a connection as we would if the loved one were simply missing instead of deceased. This seeking behavior may look like restlessness in the grieving person and, thus, looks very different from the other response to loss, which attachment theory calls despair. Despair, or giving up on the idea of trying to find the deceased loved one, accompanies an increase in inflammation in the body.[4] The experience of despair during grieving feels like a heavy blanket on us. The word *grief* is derived from the Latin verb *gravāre*, meaning "to burden," and *gravis*, which means "heavy." The same root forms the word *gravity*. We feel held down to Earth, the paralysis that results from our beloved being ripped away from us.

The theory of sickness behavior suggests inflammation contributes to this change in our thoughts, feelings, and actions from protest to despair.[5] The function of experiencing despair is that it means we stop our impossible search for our loved one, and we give up on expecting them to return, although this may be at a subconscious level. We do not think of despair as having a function, because there seems like no point to our social withdrawal and heavy limbs. We may have no motivation to seek out our deceased loved one, but we have no motivation to do much of anything else either. However, if grieving is seen as a process, we might view the cycle of protest, and then despair, as updating our expectations about the deceased loved one—coming to accept that they are gone, that we live in a painful new reality. When the process works, perhaps as it evolved to function, this acceptance of their permanent absence will enable us to establish equilibrium in our new reality, because we are no longer protesting an absence or seeking a person who will not return. Eventually, for most of us, the recognition and acceptance that our loved one is truly gone allows the return of motivation to seek out living loved ones, to engage in meaningful activities. Our heart rate, blood pressure, and inflammation return to normal baseline functioning as well.

In the throes of grieving, the people I interview often describe little interest in seeing their friends or going to social events. Knowing about the theory of sickness behavior, knowing that this social withdrawal may be our body's response to the injury of losing a loved one, might make us feel more normal. That is not the end of the story, however. There is not a single stage of protest and then a single stage of despair, and then acceptance. People cycle again and again through this process of protesting the loss and then feeling despair over the loved one's absence. It takes a very long time to learn that our one-and-only is truly gone, and we must relearn it with every new season, every new milestone. Some of us face it every single morning when we see the empty place in the bed next to us. *Where are they?! Oh, they have died . . .* again and again, like a cruel storyline.

Most of us have resilient physiology, a body that can support us through the cycling rage of protest and pit of despair. But for those of us who are struggling to adapt, over the longer term, inflammation can become chronic, or we can develop habits of isolation that may persist even after our inflammation has receded. And thus, testing out what social interaction feels like in our new world as a bereaved person becomes even more important, to reduce the long-term avoidance that accompanies prolonged grief. Returning to the world of our living loved ones may be a slow upward spiral. It may not feel great to be among people who are happily living their lives and do not recognize that we still carry the absence of someone who should be in our midst. To restore a meaningful life often means continuing to love the people who are in our material world while transforming our inner relationship with the one who has died. Our immune system tries its best to restore homeostasis (the state steady of internal conditions) during this process. But changes in the immune system may get in the way of adaptation if flooding our body and brain with inflammation reduces enjoyment of our social world and encourages us to curl up in bed and shut it all out.

If I Knew Then What I Know Now

Depression crept over me after my MS diagnosis. My life is not a well-designed scientific study, so I cannot know what part the grief over the death of my mother played in my depression and what part was the grief over the diagnosis that robbed me of my certainty about my future, my health, and my confidence in a sound body and mind. Some part of my depression was likely caused by the MS-related increase in inflammation, as depression is a common symptom in those with MS. But all of these things (mother, future, health, and confidence) were parts of me that were ripped away in my mid-twenties. I did not know what to expect from day to day when I awoke: Would I be mentally foggy? Would I have to sit down while I peeled potatoes, or be dropped at the door of the movie theater because I was too tired to walk from the parking lot? Or look like a hypochondriac when I had overwhelming fatigue on one day and then I could do everything on my to-do list the following day?

I had a diagnosis to point to, so I did not bear the even worse experience of having depression for "no reason." Honestly, having a diagnosis was a blessing. Our culture does not understand depression and how it works, but if you give it a cause in the immune system, or a neurobiological underpinning, suddenly those around you are sympathetic, and your despair is understandable, it is noble. As awful as I felt, at least I knew why. I shudder to think how others manage when their depression is not legitimized, when their grief is disenfranchised.

But I also felt despair. I woke every single morning with the thought *I am going to die.* Although this certainly was true, I knew rationally that it was not going to be from a shortened life-span due to MS. It was unlikely for this chronic disease of relapse and remission. Yet the thought was there every morning in my mind, unbidden.

Certainly coming face-to-face with death through hospice care for my mother instilled an awareness of the tenuous nature of life and the fragility of the human body. But the fact that I lived in a physical body had suddenly become unbearably apparent, because my physical body intruded on my every day. To say that I had believed up until that point that my body was there just to carry my brain around is not a joke. Sure, I occasionally had experienced through my body the pleasure of sex or chocolate, but overall, I paid no attention to it whatsoever. My absorption in school-related thoughts did not require having a stomach or muscle tone, and they were firmly outside of my focus. After I developed MS, sometimes I could not concentrate on what I was reading because of the pains in my arms; I could not haul my fatigued self to my desk to run statistical analyses; and I could not get through the mental fog to write a compelling justification for a research grant. It was overwhelming. Suddenly my ability to think and reason was being thwarted by my own ungrateful body. I had never been aware, deep in my core, that my life was dependent on my physical body, and that my physical body was finite. I despaired.

Despair contains a kernel of what adaptation means—to accept that a loved one is gone, that a diagnosis is real, that these events have affected your life forever. In this way, despair is a step on the path of grieving, further along than disbelief or protest. But despair, in all of its realism, leaves out an important part of the integration of grief into life: Although the loss has happened and there is nothing you can do to change the outcome, despair suggests that you will always feel as awful as you do now. Despair leaves out the possibility that all of these things are true and you can also restore a meaningful life. Despair leaves out hope.

How Grief Gets Under the Skin

Psychoneuroimmunology has worked out one pathway from a stressful life event, like bereavement, to increased inflammation and then to depression. Inflammation, or pro-inflammatory proteins, affect the molecular pathway that enables our body to produce serotonin.[6] From a physiological perspective, this is why a loss event in the world, and our inflammatory stress response to that event, can affect our serotonin levels, and that can lead to feeling despair. We know that serotonin regulates mood, that we feel more anxious or depressed when serotonin is not available to the receptors in our brain. That means the lack of serotonin availability leads our brain to conclude that the future is bleak, that we are helpless to do anything about it, and our best strategy is to give up.

Our body cannot produce serotonin without getting a specific protein from our diet, a protein called tryptophan. I once worked on a study in which we gave fasting people a drink that temporarily deprived them of tryptophan. Without this source protein, the body cannot make serotonin. Participants experienced depression within hours (which was quickly reversed with a different type of protein drink). We know that inflammatory proteins, especially interleukin-6 (IL-6), affect this same pathway from protein to available serotonin. Inflammatory proteins decrease the ability of the body to make serotonin and create some toxic byproducts as well. The outcome of inflammation affecting this pathway can be depression, as sometimes seen in chronic stressors like bereavement, and chronic illnesses like MS.[7]

It might sound strange, but I was one of the lucky ones with MS and depression. I had enormous resources at the time of my diagnosis. I had a supportive wife, top-notch medical care, access to mental health care, extensive education about mental health and treatment, family with experience of MS, and, most of the time, intelligence and genetically conveyed optimism. Even with all of this, my depression after the MS diagnosis meant that I could suck the energy out of a room with my mood. If I knew then what I know now, I would have seen that the inflammation my immune system spewed out varied from day to day and cyclically across each month, and I might have seen the connection between these variations and the variations in my mood and energy. But I had not yet become knowledgeable in the ways of the immune system, nor had I begun a regular meditation practice.

In the second half of this book, I share the lessons I have learned over the years of how to respond to my despair. I came to deeply understand that our mood is embodied, that moving our body can be a lever to engage motivation, but also that exerting our body can fatigue us if we fail to listen to its feedback. Most of all, I learned to expand my moment-to-moment awareness to include my body. The laser focus of my mind on work ignored the here, the now, and the comforting closeness of the world around me that broader awareness enabled. The spotlight of concentration is distinct from the floodlight of awareness. I thought mindfulness was synonymous only with the mind, but I still needed to learn that mind-fulness is also wholeheartedness—being aware with your whole being: mind, body, and soul.

The Low Likelihood of Mortality During Bereavement

If you have experienced the death of a loved one, my message is *not* that you will get MS or die of a heart attack. The number of people who are diagnosed with these diseases each year across the population is very low, and so a slightly increased rate of diagnoses among bereaved people is still a very small number of people. The overall increased mortality risk is a statistical average across millions of people, and the statistical average is driven by a small percentage who do particularly poorly. Most people have some physical symptoms during bereavement, but those symptoms are usually a sign of physiological adjustment, and not an indication that something has irreparably broken in their heart or immune system. Grief is a normal human process, an acute stress that will likely abate in time. But when it does not, when intense grief does not change over time, it can set the stage for persistent inflammation, and that persistent inflammation can be bad for multiple diseases.

Is MS more likely after the death of a loved one? In a study of 1.8 million people, Danish epidemiologist Nete Munk Nielsen and colleagues assessed whether there was any increased risk of developing MS after the death of a child or a spouse or after a divorce. They found no association between these stresses and MS risk. In 2020, looking at a more comprehensive set of ten different losses as possible contributing factors to developing MS, Xia Jiang and colleagues at Karolinska Institute in Sweden found compelling evidence for a link between these major life events and risk of MS: a 17 to 30 percent increased risk. Women were affected more than men, and events that happened within a person's past five years had the most significant effect on MS development. So I would say the jury is still out on this question. But again, I want to emphasize the very low likelihood of developing MS, even if the stress of bereavement might be a contributing factor. In a study of 30,299 Danes who experienced the death of a child, only 85 of them developed MS.

Rather than adding health anxiety as another stress to the stressful experience of bereavement, my purpose here is only to make you aware that we should all consider the increased risk, the same way we would recognize a sports injury may increase the risk for arthritis in that joint. Doctors might recommend physical therapy exercises for a person with a sports injury in order to support the injured joint. Similarly, we can benefit from learning to comfort ourselves during waves of grief and learning relaxation techniques to help soothe our grieving body.

Also, if a medical illness does develop, it does not mean that bereavement stress caused it. We cannot know the many factors that contribute to an illness. Bereavement stress is more likely to reveal an existing vulnerability or accelerate a process that had already begun. For example, if a person is diagnosed with dementia after the death of a spouse, it is likely that the process of cognitive impairment was already underway, even if symptoms were not yet apparent. Bereavement may reveal an illness because our capacity to compensate is reduced, or because the increased grief stress adds inflammation, adrenaline, and cortisol to a body and brain already burdened.

However, if a loved one has recently died, this is the right time to see a doctor. When we have been caring for a terminally ill loved one, or dealing with all the administrative tasks after the death of a family member, we may not be getting our annual physical, seeing our dentist, or scheduling our mammogram or colonoscopy. Now is the time to care for *our own health*, or to remind our bereaved friends and family to have their blood pressure tested and their moles checked. Bereavement stress is not unique—no special medical tests are needed to see its impact on the body. For example, a regular doctor visit resulted in Lizzie's pneumonia diagnosis. Antibiotics worked equally well whether or not grieving contributed to reduced immunity and resulted in her infection. We just need to seek out medical support for the grieving body.

In addition, as with all chronic conditions, behavioral stress management strategies can improve our physiology, as we will see in the next chapter. Bereavement simply affects us in the way many other risk factors might, and having a checkup during acute grief and again after six months to a year, when acute grief may have begun to abate, is warranted. Regular preventive care, with our primary-care doctor or nurse practitioner, is a sufficient response. We just need to remember to attend to it during a time when we may have difficulty getting regular things done.

CHAPTER 3

The Endocrine System

A friend of mine in her mid-forties described to me the painful wrenching of her heart in the weeks after witnessing the sudden cardiac death of her husband. With no warning, her husband had collapsed in their living room, and was just gone. Her whole life with him was torn away in a moment, from the happy home they had with their two teenage sons. In the weeks that followed, she described heartache, a literal ache in her chest, like a boulder being painfully pulled out of her chest, a sensation of having an energy ripped away and released from her. When I expressed concern about whether it was her heart, she was clear that it did not feel like chest pains or a racing heart or palpitations.

I had my own bizarre twinkling sensations in my arms after the death of my mother, which turned out to be "real." Of course, they were real whether or not a physician could explain their cause to me. I have been studying the brain-body connection for too long to ignore what my friend described, and I have heard about heartache from too many bereaved people to dismiss it. We have an actual word in our language for this experience, and I do not believe it is metaphorical. No research has investigated the cause for this sensation. But a recent study investigated where in the body bereaved people describe physical sensations following the death of a loved one.[1] Of participants, 90 percent reported that they experienced grief sensations in the chest, followed by the stomach (63 percent), head (58 percent), and neck or throat (54 percent).

Although research has not yet investigated the cause, I have a mechanistic theory based on relevant evidence. From animal research, we know that after the loss of a pair-bonded mate or offspring, there is less oxytocin bound to receptors in the brain.[2] Oxytocin is a hormone associated with nursing and lactation, but in recent years, its role in attachment relationships and social interactions has come to light. What many people do not know is that oxytocin receptors are densely represented in the heart as well as the brain. Oxytocin has a protective function in the heart. But after the separation from their pups, mouse mothers develop more oxytocin receptors in the heart acutely and then show a loss of those receptors over time.[3] There is no study connecting these changes to oxytocin receptors and physical sensations of heartache in humans, but I can nonetheless speculate that physical changes occurring in the heart could be experienced the way my friend described. We know hormonal (or endocrine) changes occur in bereavement, and we are only just beginning to understand the impact.

What Are Hormones Anyway?

Our endocrine system is made up of several organs, including our pituitary, adrenal, and pineal glands, our thymus, our thyroid, and our pancreas, and these endocrine organs release hormones. A hormone, like oxytocin or cortisol or estrogen, is a chemical messenger released in one part of our body and then carried through the bloodstream to affect tissues and organs in another part of our body. In this way, a hormone is different from a neurotransmitter, although a neurotransmitter is also a chemical messenger. A neurotransmitter, like serotonin, affects the receptors only in the location where it is released, across the tiny synapse space between the neurons. Cortisol is one of the best known hormones, key to the stress response in bereavement. When cortisol is released by our adrenal glands, it affects many different tissues and organs all over our body, including the hypothalamus in our brain and our pituitary gland, and cortisol controls the inflammatory markers our immune cells release.

Because the hypothalamus is in our brain, the unified system of the hypothalamic-pituitary-adrenal (HPA) axis enables information perceived and processed by our brain to affect our body through the production of cortisol. Cortisol's role in stress is to send information about our brain's priorities throughout our body. For example, cortisol controls our metabolism, or how our body gets energy from the foods we eat, including regulating our blood sugar levels. If we need more energy to deal with a stressor, cortisol increases the glucose available to our muscles and brain. Cortisol also controls our sleep–wake cycle, since stressors may require us to be more aware of our environment.

The Australian bereavement study by Buckley and Tofler described in chapter 1, which discovered heart rate and blood pressure are higher during the first six months after the death of a loved one, also investigated cortisol levels.[4] The researchers found that morning levels of cortisol were higher in the acute grief group than the nonbereaved group. This is a highly replicated finding in grief research; numerous groups have found elevated cortisol during the stress of bereavement. Even in pair-bonded animals, like prairie voles, who mate for life, the loss of their partner causes a significant increase in stress hormones (called corticosterone in rodents).[5]

For anyone who has experienced acute grief, learning that cortisol floods the body cannot be surprising, given how stressful bereavement feels in the early days, weeks, and months. But unlike the data showing cardiovascular measures returning to normal levels by six months, making bereaved and nonbereaved groups indistinguishable at that point, Buckley and Tofler found morning cortisol remained high. In considering why illness and death may increase in bereavement, I find it intriguing that cortisol levels may not return to baseline for months, and for some, cortisol dysregulation may ensue. It is likely that this was the case for me after my mom died, although, as I will describe next, this is hard to figure out for an individual person.

The Many Ways Cortisol Can Become Dysregulated

Cortisol is regulated in a couple of different ways. First, our body uses a feedback loop, like a thermostat, to determine how much cortisol to make. When cortisol is produced by our adrenal glands, our hypothalamus and our pituitary gland pick it up. Like a thermostat, our HPA axis has a set point. If the amount of cortisol our hypothalamus and pituitary are detecting is higher than that set point, they stop making the chemical messengers that would lead to more cortisol production, like shutting off the heat once the temperature in the room reaches the level set by the thermostat. If our hypothalamus or pituitary is not detecting enough cortisol, they increase their chemical messengers, which are picked up by our adrenal glands, and our adrenal glands produce more cortisol. The whole process is an elegant orchestration.

A second regulation of our cortisol release occurs in a regular pattern across the day. Cortisol has a diurnal pattern, which means that our cortisol is lowest when we wake up in the morning, and then it shoots up in the first thirty minutes, giving us the energy to get out of bed and deal with our day. In fact, the rise in our cortisol is higher on workday mornings than on weekends to give us the energy to deal with the demands of our job.

Our brain is always trying to predict what our body will need to cope with what we encounter. After its normal rise after we get of bed, our level of cortisol slowly decreases across the day, enabling us to go to sleep in the late evening. This pattern repeats every day, with a sharp increase in cortisol in the morning and a decrease across the day, creating a long slope from high levels to low levels. Short bursts of cortisol happen in response to momentary stresses, and the bursts are overlaid on this general pattern throughout the day. The rhythm of our lives is reflected in our cortisol release and regulation.

The complexity of this system of feedback loops and diurnal patterns means there are many opportunities for the system to be affected. When typical stresses of day-to-day life occur, cortisol provides helpful information for our body to shift priorities toward spending energy on coping and then shift back to its usual priorities, like digesting food or rebuilding muscle. In order to measure cortisol response and recovery in the laboratory, we ask study participants to do a standardized task, called the Trier Social Stress Task (TSST), in which people are told they are going to have to give a talk in front of some not-very-kind judges and do math problems out loud. For most people, we see a brief rise in cortisol as they prepare and give the talk, and then it falls back to their baseline in about forty-five minutes.

But some chronically stressful situations overwhelm the HPA axis and change our set point, because the elastic resilience of the system has been depleted. The stress of bereavement can become a chronic stress, accompanied by high levels of cortisol production for a long period of time. Chronic stress becomes the background landscape for how the body responds to acute stressors. Under chronic stress, the body does not make much of a cortisol response when it is faced with an acute stressor like giving a talk. Psychologists describe this as a blunted response.

In my own laboratory and in others, we have seen this blunted response in those who have experienced loss, including breaking up with a romantic partner or moving away from home in the past six months or losing a loved one in the past year.[6] In the following figure, you can see the data from a study done with my former graduate student Monica Fallon. The curve of the solid line representing those who had a recent loss does not reach the height of the curve of the control group who had not experienced loss. You can imagine that not having enough energy resources might affect your performance while giving a talk or having to deal with the many tasks that accompany bereavement.

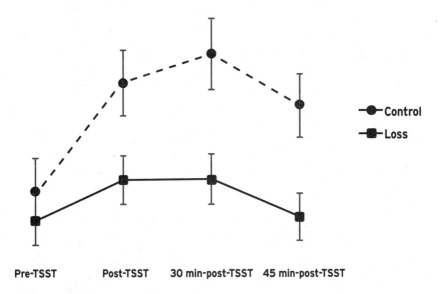

Pre-TSST   Post-TSST   30 min-post-TSST   45 min-post-TSST

Remember that in addition to responding to acute stressors, our normal diurnal rhythm of cortisol is typified by a long slope across the day, with a high level of cortisol after we awaken in the morning to meet the demands of the day, followed by a decrease, enabling us to fall asleep at night when it is low. In many bereaved people, the level of cortisol does not vary as it should across the day, with a lower-than-normal level when getting up in the morning and a higher-than-normal level of cortisol in the evening. This means little variation across the day, or a *flattened slope* compared to the expected steeper slope. Studies show that a flattened slope of cortisol production is more common in those with prolonged grief, even two to five years after a loss.[7] In other studies with non-bereaved people, a flatter diurnal slope of cortisol is associated with worse health, including increased risk of cancer progression and depression.[8] Notably, cancer progression and depression can also occur during bereavement. Although no research studies have triangulated these findings yet, I imagine someday we will discover that a flattened diurnal cortisol slope is one mechanism linking bereavement to these health outcomes.

Desensitization to High Levels of Cortisol

I mentioned that there are multiple ways our HPA axis can become dysregulated. A third type of dysregulation can occur during chronic stress, distinct from the blunted response to acute stress or the flattened diurnal slope. When our body has experienced large amounts of cortisol for too long, it may reduce its number of cortisol receptors. You can imagine how this works: If you are bombarded by someone shouting, you plug your ears eventually. Similarly, our body changes its response if it experiences continuously high levels of cortisol. Your body may find a new normal, by becoming desensitized to the high levels. Reducing the number of cortisol receptors enables our body to ignore the high levels being produced. With this reset, although high levels of cortisol are still produced, fewer cortisol receptors mean our body ignores some of it as background noise, rather than useful information.

This reduction in our cortisol receptors can have significant effects. When cortisol receptors are decreased on immune cells, inflammation can increase. This is because cortisol ordinarily decreases the production of inflammatory proteins. Have you ever put hydrocortisone cream on the inflamed skin of a bugbite? Hydrocortisone, a synthetic version of cortisol, reduces inflammation of our skin. When our immune cells detect cortisol, they typically decrease their production of inflammatory proteins. When the number of cortisol receptors decreases under chronic stress, our immune system no longer reduces production of inflammatory markers—because our immune cells do not detect the cortisol. Thus, if chronic stress creates high levels of cortisol for a long time, followed by our immune cells reducing the number of their cortisol receptors, then the information from cortisol no longer reduces the production of inflammatory proteins, and we can see one way chronic stress can lead to increased inflammation.

We know a lot about the connection between cortisol and stress, and our knowledge tells us how cortisol varies between different groups of people, like bereaved and nonbereaved. But this knowledge does not tell us much about a single person's level. Bereaved people *on average* have low cortisol in the morning, but I cannot tell you whether any particular bereaved person who walks into my laboratory (or a doctor's office) will have low cortisol in the morning. This is similar to knowing that men are taller than women, overall, but not knowing if a specific woman will be short or tall. Medical conditions arising from too much cortisol (like Cushing's disease) or too little cortisol (Addison's disease) show circulating cortisol levels at extreme highs or lows. These conditions can be detected in individuals through a blood draw in a lab. However, the amount of change in cortisol due to life stress is much more subtle, because of normal day-to-day variation of cortisol. Measuring the cortisol or inflammatory markers of an individual at one time point does not tell us very much about their state of chronic stress, unless their cortisol or inflammatory markers are far outside the normal range.

This can feel frustrating for a patient. For example, as well as I understand chronic stress and cortisol, I cannot tell you if my cortisol levels are related to my fatigue. My own values from a blood test would simply fall within the normal range that the test is designed to convey. The difference between my cortisol levels on the day I struggle to get out of bed may be different from my cortisol levels on an average day, but medical tests are not designed to compare one day to another in the same person. So although we know a great deal about cortisol dysregulation in laboratory studies of stress in general, we will have to wait for precision medicine to develop tests that can tell us about the importance of our individual variation in cortisol.

Cognitive Stress Theory

Attachment theory serves as a lens to explain how the death of a loved one creates a physical blow, because a part of "us" that developed through bonding disappears. Attachment theory answers many of the questions about why grieving is so painful and takes so long. But in the 1980s, another theory developed to explain the difficulties produced by stressful life events. The *cognitive stress theory*, proposed by psychologists Richard Lazarus and Susan Folkman, has also been applied to the death of a loved one.[9] Bereavement consistently ranks as the most stressful of all life events. Cognitive stress theory highlights bereavement as an *added* demand for our system to cope with as opposed to a *subtracted* functional part of us, as attachment theory points out. Both cognitive stress theory and attachment theory have valuable insights for understanding grief. Looking at bereavement through a lens of stress and coping, much can be gleaned by considering how our body responds similarly to other types of stressors, all of which push and pull on our endocrine system, which houses our stress hormones.

It may help to first delineate what constitutes psychological stress, using the explanation by Folkman and Lazarus. They define stress much like this: We experience stress when we perceive that the demands of the world exceed our resources to cope. I imagine an old-timey balance scale when I think of this definition, with demands on the left pan and resources on the right pan, and stress as the state when the scale tips toward the demands side. By thinking about stress in this way, we can consider multiple methods with which to improve our situation when we feel stressed. For example, we might be able to decrease the demands on us. If we have always volunteered for our child's school, bereavement may be a time when we decide to take a break from that role. Some demands of the world are nonnegotiable, such as a job, and this points to the importance of bereavement leave. But there may also be creative ways to reduce how much we do when we have fewer mental and emotional resources. Sometimes that means putting something off until later and thereby spacing out what is on our plate, or even realizing that a demand is not worth it in light of the realization that life is short and time is precious.

Thinking of a balance between demands and resources, we can also intervene by trying to increase our resources for coping. This can mean giving ourselves the opportunity to sleep for eight hours or taking time to exercise, because these both add to our physical stores, which help us cope with the demands of the world. Increasing our resources can simply mean asking for help, with organizing and donating our loved one's clothing or taking our child to a dentist appointment. Sometimes increasing our resources means giving time and attention to philosophical questions that may plague us, like how to live a meaningful life, or how the experience of death has changed our relationship with God or the universe. Taking time to read books by others who have tackled these questions, or reaching out to a pastor or meditation teacher to have a conversation, can be time well spent, giving us new resources with which to cope. Increasing our resources requires creativity: What would it feel like if you gave yourself more bandwidth to cope with the continuous onslaught of thoughts and feelings of grief? Or what resources would increase your energy or ability to focus?

When we cannot change the demands of the world, we may be able to increase our resources, and when we cannot change our resources, we may be able to reduce the demands on us. But an aspect of Folkman and Lazarus's definition of stress that may have escaped your notice is the word *perception*. Psychological stress arises when our perception of the demands of the world exceeds our perception of the resources we have to meet them. Even when the demands or resources do not change, sometimes we can change our perceptions of them. This can happen by reducing our expectations. Perhaps what feels like a demand is actually a request, and we could negotiate what it means to fulfill that demand. Often we carry around what we "should" do, and what we should do can create a lot of stress. Who defines *should*? The word implies someone has expectations—are those expectations of ourselves, and the world would not actually fall apart if we ate cereal for dinner all week? Or sometimes the "should" is an expectation of someone else—and do we care if our child's teacher sees that they have worn the same clothes two days in a row? Changing our perception of the demands on us requires us to value reducing our stress as a worthwhile outcome, an outcome more worthwhile than what others may say.

I cannot emphasize this enough—this is a change in recognizing what we value, even in the face of others' judgment, and it takes great courage. In order to find this courage, we must become convinced that enduring the stress associated with particular demands is not worth it. It requires valuing the relief we feel when we reject those demands, or valuing the ease we experience when we ignore those self-imposed demands, in order to make time for us to feel relaxed. The feelings of relief and ease when we let go of demands echo in our endocrine system, and thereby affect our physical reaction to the stress of loss as well as our emotional stress reaction.

Changing our perception of our resources, and not just our demands, can also reduce our experience of stress. Just focusing on the fact that we have resources—bringing to mind again and again the fact that we have coped with difficult things in the past—can reduce our experience of stress. Increasing our awareness of our resources, even without increasing the resources themselves, can be soothing or encouraging. Focusing on existing assets, like imagining a friend who listens or an upcoming long weekend, helps loosen tension in our body or bring down our heart rate. Grieving is already physically stressful, stressful enough to cause physical illness. Coming to understand how our thoughts and our evaluation of our situation can amplify or buffer our stress experience is valuable during this time of higher physical challenges.

My Own Chronic Stress

My mother's cancer progression and terminal illness across my adolescence and early adulthood was enormously stressful for me. My mother was a brilliant woman committed to social justice for women and children, and a tremendous supporter of the arts. When I mention my mother dying when I was twenty-six, I think most people imagine I felt great yearning for my mother, that I wanted her back. This is how we expect people to feel after the death of their mother, and yearning is a common, painful experience of grief. But I did not feel bereft. Instead, I actually felt great relief. People often feel relief after the death of a loved one, because a terminally ill loved one can suffer for a long time. Their physical body is no longer able to go on living, but dying takes time and causes pain and distress. I felt this kind of relief because my mother had been battling cancer for thirteen years, with difficult rounds of chemotherapy, great uncertainty about her prognosis and when the next recurrence would strike, and periods of deep depression. I felt relief because she was no longer suffering, but this was not the only reason I felt relief.

Relationships are complicated. My relationship with my mother was fraught, and managing our interactions caused me great stress, anxiety, and a feeling of helplessness. I struggle to write this even a quarter century after she died, but at the time, my mother's death felt like a solution for me. I could not figure out how to make us both happy, but I desperately wanted her to be happy, and I wanted my own happiness as well. My mother's life, even before she was my mother, was colored by clinical depression, worry, and guilt. She told me that when my sister and I were little, before we started school, was the happiest she had ever been. I think she felt a sense of purpose, reinforced by our needing her every day, for everything. She was a wonderful mother to us as small children, a full-time caregiver and homemaker. Our needs were her needs, and her needs were fulfilled by fulfilling ours. But as we grew older, our needs changed. We wanted friends and other adult influences in our lives. We wanted to explore our world beyond her view. Her fulfillment through the symbiosis of mother and child no longer soothed her mental health demons, and they roared back.

I recognize now that conflict with my mother was partly the normal stretching and snapping of the attachment bonds that held us in the mother-daughter relationship. I wanted to defy the strict control she could not give up without losing a feeling of closeness, and I also could not bear it when my separation from her caused her distress. I happened to be born with, and developed through my early life experiences, a brain that is very vigilant and reactive to the feelings of those around me. This is also a gendered experience—there is a continuous training of girls to focus on others' feelings and how to accommodate them. But I was particularly sensitive, and I experienced no space between my perception of others' feelings and the idea that I must respond to fix them. I was never sure what I wanted for myself, from an internal compass, and what I wanted because she wanted it. Her death, I thought, would solve all of that conflict.

You would think that wanting my mother to be happy defines compassion. And my impulse, my basic desire for her happiness, was a good thing. I had empathy, in the sense that I was acutely aware of how she was feeling. But I had no skillful way to evaluate what would actually make her happy, or how to view the happiness of us, rather than of her or me. I believed that I was the source of her happiness or unhappiness, but I no longer believe this is true. I share all this introspection about my childhood to explain how the relationship between my inner mental world and my stress physiology developed. Always vigilant to the relationships around me, I ignored parts of my internal experience that conflicted with what others needed. I overruled my feeling tired, or disrupted, if someone needed something. One cannot ignore a part of one's internal experience and live a healthy life. If we can think of grieving as a form of healing, I had a lot to learn about how my relationship with my mother, and with others after her death, was related to my anxiety, my hyperactivity and yet utter fatigue, and my chronic illness.

The Stress of Restoring a Meaningful Life

When we think about what is stressful during bereavement, most of us think of the death itself, or the waves of grief that overcome us, the intrusive thoughts we cannot get out of our head. But an insight from psychologists Maggie Stroebe and Henk Schut at the Utrecht University in the Netherlands led them in the 1990s to develop the *dual process model* of coping with bereavement.[10] This is now a well-regarded model, used by grief researchers and clinicians alike. In this model, Stroebe and Schut identified a missing element from prior models of bereavement coping. In addition to loss-related stressors, bereaved people must deal with restoration-related stressors. Not only must they learn to carry the absence of their loved one and learn to manage the fact that they are now people who have waves of grief (loss-oriented stressors); they must also figure out how to live a meaningful life without the loved one (restoration-oriented stressors). Although there are many universal aspects to grieving, it is a very individual process—each person must figure out what their loss means for them, how to reestablish their own values and goals, and what will lead them to experience joy, creativity, compassion, and pride. These positive feelings can coexist with the painful feelings our human experience contains, including grief, yearning, sadness, anger, and anxiety. All feelings are important, and the sign of mental health is that we are able to allow these feelings when they occur, and discover ways to respond to them that forward and improve our lives.

I am fond of the word *restoration* that Stroebe and Schut chose. I often think of restoring an old house as a metaphor for what needs to happen in the wake of a loved one's death. The structure of the house is still there, but it probably needs new paint and new curtains, and maybe even bigger changes like upgraded electric and plumbing. You are not moving away from or giving up on your old life; you are simply restoring it in ways that will make it a comfortable and enjoyable place to live now. I also needed to restore my life after my mother died, and then again years later, after I got divorced and my father died. The waves of grief we experience in acute grief tend to occur with less frequency and intensity over time, and we learn to cope with the stress of loss. So, too, we must learn to cope with the stress of restoration, until it becomes just one part of our everyday experience, present but not overwhelming.

But here's the thing: If you have ever owned a house, you know that house maintenance is never done. We discover new leaks, or appliances age and need to be replaced. Creating a meaningful life is always a work in progress, for as long as we draw breath. In that way, grieving never ends, because the need to restore our life will continue as we discover new losses that the death of a loved one creates. For example, years after my father died, having a new partner means I have new grief that they never knew each other.

The challenge for me, in learning to restore my life after the death of my mother, was I assumed that without her in my external world, my internal world would become less stressful. In fact, when we have a bond with someone, they continue to live on in an internal representation, sort of an avatar in our mind. Transforming these continuing bonds may be necessary in order to move forward in our life. For some people, the loved one simply appears less often in their daily mental activity, although they can be recalled. But for others, the patterns of interaction we learned from one important relationship continue to shape the way we interact in new relationships. Our expectations of how others will react to us results from many lived experiences of moments with our first bonded loved one. But in fact, any relationship is only 50 percent us and 50 percent the other person (plus some alchemy of the relationship itself). Continuing to negotiate with the new person in the way we did with someone before can lead to similar behavioral patterns repeating themselves in a new relationship, a self-fulfilling prophecy. And so, for me, I continued to have only a little sense of myself even after my mother's death. In an argument, I often could not keep track of what I believed once I heard the other person's side, much as had happened when I argued with my mom.

My stress level did not decrease over time, as I lived my new reality after my mother's death. But in fact, our physical stress response can be used as information, to adjust our ongoing behavior in coping with the world. That stress response is our body's way of communicating what it is taxed by and what it needs to regain balance. On the other hand, if we are not attending to our stress response, ignoring it entirely, or if we are not skillful at using the information from our body to respond differently to the demands of our world or our resources for coping with them, our stress response will continue. And the stress will continue to tax our physical resources, in the form of cortisol and other hormones, and for some of us, eventually affect our health.

Bereavement Is a Health Disparity

Our bond with a beloved person means that we function in the world as an "us." The presence of the loved one regulates our own physiology, well below the surface of consciousness; so after their death, a motivation to seek them out may emerge from powerful chemicals created in our own brain and body, including dopamine, opioids, oxytocin, and cortisol. But what if this expanded view of grief is still too narrow? What about collective grief, for those who are bonded with a community of people? What if "us" is a whole group of people, both known and unknown, and our grief is inextricably linked with theirs?

Most of us have never thought of it this way, but bereavement is a health disparity. Mortality rates are different across different racial and ethnic communities, which means rates of bereavement differ as well. With COVID, we witnessed brutally higher mortality rates in Black, Indigenous, and people of color (BIPOC) communities. Black children were two and a half times more likely to lose their caregiver than white children, and Native American children were four times more likely. Of course, bereavement was a health disparity pre-pandemic as well: People from marginalized backgrounds lose more family members and community members, and they lose them at younger ages in their own development. Michelle Chang and Ted Robles at UCLA describe this bereavement disparity as "too much" and "too soon."[11] Because marginalized people may have fewer resources to begin with, the loss of a parent or an adult child or even a neighbor may have a greater proportional impact on their lives.

The stress of grieving multiple losses affects our endocrine system, which influences every other organ system in our body. The medical impact of multiple bereavement experiences in the Black community has been an area of careful scientific study. Tené Lewis, a psychologist at Emory University, conducted a longitudinal study of African American women in midlife, examining the role of many types of stressors on their heart health. She discovered that African American women who experienced three or more upsetting deaths across the prior twelve years had thicker carotid arteries, compared to Black women with two or fewer losses, and compared to white women with three or more upsetting losses.[12] Greater artery thickness is a sign of early atherosclerosis, before a person is symptomatic, increasing the risk of heart attack and stroke. Thus, bereavement in midlife may be one pathway through which race/ethnicity influences cardiovascular risk for African American women. The importance of this information cannot be underestimated—aggressive medical care can reduce the impact of atherosclerosis when caught early.

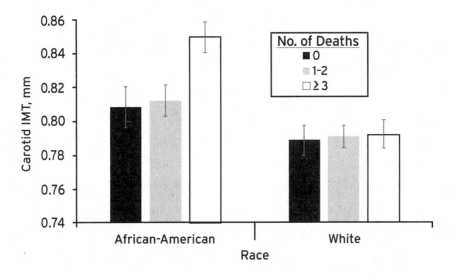

The Centers for Disease Control and Prevention (CDC) reports much lower life expectancy in the Black community, at 71.8 years in 2020 compared to 77.6 for white Americans.[13] Think about the difference of 5.8 years—how different it would be to lose your grandparents when you are six instead of when you are almost twelve. Although the gap between Black and white life expectancy had been narrowing in the past three decades, COVID wiped out the improvement, widening the gap again. The stress of multiple losses can result in lifelong health issues, and the timing in one's life matters. The impact of grief on the developing body, brain, and spirit needs to be addressed by bringing down the mortality rate but also by providing culturally appropriate bereavement support for those who need it most.

Celebrating Día de los Muertos

When I talked with Director of Community Engagement Lance Meeks in his office at Goodwill Industries of Southern Arizona, I was aware that, in addition to our conversation, he was continuously monitoring six or more situations affecting opportunity youth at the Goodwill Youth Reengagement Centers (REC). Opportunity youth are young people aged sixteen to twenty-four who are disconnected from work or school, for one reason or another. Often these reasons have to do with a system they have been put through, like the foster care system or the juvenile justice system. The reengagement centers offer a safe space to explore educational and career opportunities, get help with job applications and interview skills, progress toward a GED, and participate in community events and service projects. As I watched Meeks interact with the young people around their large conference table, or even play video games with them occasionally, it was clear that he saw opportunity first in each one, and considered their background secondarily.

This perspective led Meeks, when he considered how to make sure the reengagement centers were safe places, to engage in initial conversations with each young person about their life experience. From these questions it became clear that the youth coming in had experienced a tremendous amount of grief and loss in their lives, that the death or loss of someone close to them was often when their lives went off track, and that their grief had never been addressed. Meeks recognized this as a chance to partner with the young people, along with Tu Nidito, a local Tucson bereavement support organization, and End of Life Care Partnership, at United Way of Tucson and Southern Arizona. Together they created bereavement support groups that take place on-site at the reengagement centers, in their safe space.

I could feel the specialness of the Youth REC as soon as I visited. I offered to talk with members of Goodwill's Youth Council about celebrating Día de los Muertos, recognizing it might be a culturally relevant way of approaching grief for some of the youth from Mexican backgrounds, and because Tucson has had a large public celebration for all-comers for many years. The youth described the rituals for celebrating or mourning their deceased loved ones in their own families, from wearing black to lighting candles to releasing balloons, fascinated that so many different memorializations existed. An important Mexican tradition is creating an *ofrenda*, an altar and offerings for the departed, and the youth decided they wanted to create one.

They planned a day of remembrance, including face painting and making paper flowers, eating favorite foods, and listening to music their loved ones enjoyed. They made memorial posters for each person, and we walked together to place them in a designated spot in the neighborhood. One youth, Myleigha Truitt, wrote, "We were able to be vulnerable amongst each other by sharing our loved ones and even singing. We had a moment of silence and ended by placing our items on the designated wall to share with the community."

The youth have re-created the Día de los Muertos celebration for several years now. When I talked with them, I became aware of the fact that prior to their experience at the Goodwill reengagement centers, many of them had never told anyone about the grief they were feeling, the losses they had experienced. I cannot begin to imagine never talking about losing my mom, or never hearing what another human being thought about grief and loss. I genuinely cannot imagine how I would have figured out what was going on inside of me, to even recognize that these were feelings of grief. The transformation of the young people through acknowledging their grief is tremendous. They explained that when no one notices their sadness, eventually those emotions turn into anger and rage.

One young man shared with me that his uncle had died by police violence a few days before one of the grief support groups started, and he was really stressed about coping with it. In the group, suddenly he was given multiple ways to cope, from learning to breathe as a way to calm himself to writing letters to his uncle, saving them, and being able to read them later. Many of the youth described how hard it was to open up in the groups. They were amazed when they did, because they were never thought of as weak, even when they cried. Most touching to me, they talked about how powerful it is to be in a room full of young people with the same experience. While sitting and sharing something so vulnerable, they automatically form a bond. Everyone understands. Outside of the support group, they recognize if one of them needs help and whether they have someone to reach out to. Even in the circumstances of unbearable loss, with the accompanying increase in stress hormones, grief can be buffered when we tip the balance scale by filling the pan with resources.

Striving for a World Where No One Grieves Alone

The fact that I had the time and support to grieve over the death of my mother and the diagnosis of MS meant I developed meaningful ways to view the world from these losses. Without the opportunity to read books on philosophy and attend meditation retreats, I might not have come to the same understanding of how to listen to my body's symptoms of stress and how to practice self-compassion, as I will talk about in chapter 10. As Breeshia Wade writes in her book *Grieving While Black*, "Our social location determines our access to time and thus to our ability to grieve and reintegrate after loss."[14]

Without these resources, I might have come to very different conclusions—that the world is indifferent, that I am not worthy of love, that I will never amount to much. Acknowledging this privilege does not mean it did not take courage to heal after shattering loss; it means that I was born with a lighter load in life before grief weighed me down. This has cemented my commitment to work toward the social justice that might allow everyone the space to grieve, the time to restore meaning, and the resources to understand how human grief works. I recognize that if others do not have the space to grieve, then everyone in our community loses out. And so, like the mission statement of Tu Nidito, I work toward a world where no one grieves alone.

CHAPTER 4

The Liver and Lungs

M y mother died in the early hours of the morning, three days after Christmas. I have no memory of returning to my childhood home that night to sleep, but I remember that my best friend was there when I got up in the morning, and we went downtown to get breakfast. In a town as small as ours, everyone knew that my mother was gravely ill. When the owner of the Mexican restaurant we chose came by our table, she said, "I'm so sorry to hear about your mom being in the hospital." I thanked her and told her that she had died the previous night. "Oh, *m'ija*," she said. "I'm so sorry. What can I get you? Anything you want."

"Can I have a beer?"

"Of course you can."

Significant losses can create two problems. The first problem for your health is the grief itself, with the accompanying physiological changes that threaten to overwhelm your heart and immune system. The second problem is your response to the grief, your strategies for coping with the pain of grief. These coping responses can become their own problem for your health. The pathway from grief to medical consequences can go on much longer than the acute stress effect, created through your health behaviors, through your attempts to cope with the stressful loss, chipping away at your body day by day. Because although having a beer for breakfast can be an appropriate coping strategy on the morning after your loved one dies, if it becomes your primary coping strategy, it will do long-term damage to your liver and may even shorten your life.

The research on how alcohol, cigarette, and drug use patterns change during bereavement shows that the majority of people who have lost a loved one do not drink or smoke more than they did before.[1] Nor do bereaved people use more than similar nonbereaved people. Although most people do not change their habits when they lose a loved one, the smaller group who do increase their use of substances is of serious concern. Bereaved men are more likely than women to report increased drinking, and the rate of clinically at-risk alcohol consumption continues to increase for the first two years following bereavement.[2] People who are bereaved due to a loved one's death by suicide, accident, or homicide are more likely to increase their use of substances compared to deaths occurring by natural causes.[3] But most importantly, the consequence of substance abuse can be fatal.

Mortality due to alcohol-related causes (including alcohol-related cardiomyopathy, gastritis, liver disease, pancreatitis, alcohol-related accidents and suicide, alcohol poisoning, and alcohol abuse) was substantially higher among bereaved parents in a very large Norwegian study across thirty years.[4] This mortality risk was higher among women, although the risk for both sexes was higher than the risk among nonbereaved people. The higher mortality rate was true regardless of whether the parent lost an adult son or daughter or a child under the age of eighteen. Even clearer evidence of this increase in alcohol-related mortality came from a thoughtful comparison the researchers investigated. Because many factors, both from one's genes and one's childhood environment, may influence alcohol abuse and alcohol-related disease, the researchers compared the bereaved parent to one of their own siblings. This analysis showed that the death of a child increased alcohol-related mortality only in the bereaved sibling, and thus over and above any risk due to genetic or familial factors.

A Big Toolkit of Coping Strategies

Human beings have a myriad coping strategies to handle stressful situations, from expressing our feelings to a supportive friend, to changing the way we view a situation to make it easier to manage, to suppressing or avoiding our feelings altogether. Research suggests that there is no wrong coping strategy, even avoidance. More important is using the best coping strategy to match the particular situation, the context, in which we find ourselves.[5] Coping includes efforts to minimize, tolerate, overcome, avoid, or control a stressor.

Frequently, our coping strategy leads to a change in how we feel, since our feelings often arise in response to a stressful or unpleasant situation. Flexibility in how we cope, or in changing what we do and how we think about a situation, gives us a wider repertoire to deal with the various new experiences we confront as grieving people.

Having a wide range of coping strategies we can deploy allows us to vary our response to the context. The context can include all sorts of dimensions, from who we hang out with to how long the situation will last to what time of day it is. How I cope with a wave of grief when I am going to bed differs from how I cope with it when I am about to go into a work meeting. I think of this as carrying around a large toolkit of coping strategies, and then looking at a situation and thinking, *Wrench? Screwdriver?* A big toolkit prevents us from using the same trusty hammer in every situation, since a hammer might not be the best tool when fixing a window. The toolkit metaphor could be translated this way: *Go for a walk? Vent to a friend? Binge watch* The Expanse? Developing an awareness of what coping tool we reach for unconsciously can help us to find better choices in how to deal with the pain of grief.

Ways of Avoiding

When we are grieving, we often wish we could shift how we are feeling, shift out of grief. In fact, most of us would like to shift our whole reality and have our loved one back, to have our life like it was before. But at a minimum, the incessant feelings (and thoughts) of grief may make us want to shut the book of life and stumble through our day/evening/weekend like a zombie, unaware that we are stumbling through our day/evening/weekend like a zombie. One of the most common ways we avoid whatever is happening in the present is through ingesting a substance that changes our state. We ingest drugs for many reasons, including legal drugs. Think of alcohol, marijuana, nicotine, prescribed painkillers, and caffeine. We use substances to increase our energy, decrease our anxiety in social situations, give us a break from work, improve our mood, help us wake up and help us get to sleep, motivate us, relax us, celebrate occasions, improve our concentration, reduce physical pain, mark ritual times in our day, blot out our thoughts or feelings, and deal with boredom. What do all of these have in common? We want to change something about our present state, and we rely on a substance we ingest to get us to that new state.

My friends and I often observe Dry January, not drinking alcohol for the first thirty-one days of the year. I engage in Dry January because although I have never had a substance use disorder, drinking alcohol can become such a habit that I do not stop to think about why I am doing it. And I think the most important thing about understanding our motivation for having a drink (or using something else) is not about the state we are trying to achieve but about the state we are trying to avoid, or undo, or shift. When the habit of pulling out a pack of cigarettes or stopping for a second cup of coffee flies completely below the radar of our awareness, we have little opportunity to learn what is working or not working in our lives, and what else we might do to shift how we are feeling.

In the book *Waking Up to What You Do*, Buddhist teacher

Diane Rizzetto describes a process through which we can become aware of our automatic habits, the responses to our situation that lead us to make the same choice over and over. For some of us, that choice is automatically picking up a glass whenever grief infiltrates our peripheral awareness. Suddenly we are answering the standard doctor's office question of "How many alcoholic drinks do you have in a week?" with more drinks than days, having never noticed that we slipped into this pattern. Rizzetto describes the opportunity to insert a pause before reaching for a bottle or cigarette pack or Starbucks and reflect for a moment on how we are feeling. Long before we might decide to do something different at the end of that pause (perhaps choosing something other than ingesting a substance to shift our state), we might allow ourselves to feel that moment, to feel our body and mind and desires and fears. Whatever we feel has value, even if only for the moment.

Often the grief we are feeling seems unbearable. Reflecting on the pause we could insert between reaching for and picking up that glass, Rizzetto says, "It can reveal the blueprint of the self-centered dream in which we view the world as we think it *should* be rather than how it *is*."[6] *My loved one* should *be here. I* should *be stronger than this. I* should *feel lucky we had each other, rather than angry they are gone. I* should *be happy to be at this party, with people who I know care about me.* An infinite number of these *should* statements are possible, and yet none of them truly matter. What matters is what you feel right now, and whether what you feel is going to dictate what you do next. Because these internal requirements for how things should be and how you should feel are a prison—unfortunately, a prison that can lead to lung cancer or liver disease.

Addiction

I am not an expert on addiction. But I know from psychological science that there is an unclear boundary between the way a substance can be used occasionally as a way to cope and the way a substance ultimately can change our brain and, consequently, our motivation.[7] What shifts on the way to addiction is often all about our view. When despair about the future or self-loathing are dominant in our mind, these looming "truths" can lead us to want to filter our consciousness through alcohol or painkillers or some other method. People with addiction view drug use as the only way to cope, while feeling helpless to come up with other strategies in difficult circumstances. Eventually, coping without having the drug on board seems impossible. We would make better choices if we believed there was a future worth the effort.

We use better coping strategies when we have resources, when we have friends we can call on to comfort us and lend us their hope, when we believe we are capable of making it in this world without our one-and-only, or even that we will make it in this world someday. Addiction does not rob us of our free will, but through the changes in our brain wrought by a drug, it affects our ongoing decision-making, distorts our thinking, and makes our use of alternative coping strategies less reliable. This can happen to anyone. When I broke my ankle in my forties after my divorce and while my father was terminally ill, I was given painkillers. I had taken painkillers after a knee surgery when I was younger, and I had no problem taking them in the prescribed manner. But this time the situation had changed. Suddenly the painkillers did not just kill the pain. They gave me comfort. They made me feel better about myself and my lonely situation. Recognizing they made me feel this way scared the heck out of me, and because I had the good fortune of many other resources in my life, I got rid of the pills and tapped into other options to deal with my loneliness, fear of failure as a person, and moments of despair. Context matters as to how a drug affects our system.

All Moments Are Not Created Equal

The idea that drugs might affect us differently during times of stress, like bereavement, has a basis in psychology research as well. Drugs with abuse potential (including alcohol, nicotine, and opioids) increase the body's production of beta-endorphins, the form of opioids we produce naturally. In fact, drugs of abuse stimulate endorphin and dopamine release in an area of our brain called the nucleus accumbens—exactly the same brain area we see activated when we yearn for a deceased loved one. The *brain opioid theory of social attachment* has a long history in animal neuroscience, and a more recent history in human neuroimaging studies. Separation from a loved one results in an abrupt decrease in our natural opioid production, and this may be part of why bereaved people describe grief as "painful." For example, when women were asked to reflect on memories of separation from a loved one, such as bereavement or breakup, brain imaging showed there was less opioid binding in the brain during this reflection period, in an area directly adjacent to the nucleus accumbens.[8] The description of grief as painful may not be just a metaphor but also a vague awareness of this physical change in our sensitivity to pain. Consequently, drug use during bereavement may be a form of self-medication to overcome this painful state.

Psychologist Mustafa al'Absi and his colleagues at the University of Minnesota Medical School study addiction.[9] Al'Absi believes a dysregulated physiological stress response might predict smoking relapse. For example, we know that periods of increased stress lead to higher rates of relapse. In a study to test this hypothesis, forty-five smokers who wanted to quit went to Al'Absi's laboratory in the first twenty-four hours of their abstinence. They were hooked up with an intravenous (IV) catheter so researchers could take small blood samples of their beta-endorphin levels throughout the study. After adjusting to the laboratory environment, the participants were asked to give a speech defending their position on a social issue or defending themselves against an accusation of shoplifting. (I would certainly find this stressful!)

Of the forty-five participants, thirty-four of them relapsed within the four-week follow-up period, which is not uncommon in quit attempts. Measures of craving, withdrawal symptoms, and distress increased significantly for all the participants after giving the speech, regardless of whether or not they went on to relapse. And overall, when they showed up for the study, the level of beta-endorphins was the same for those who relapsed and those who had successfully abstained. But notably, relapsers had significantly lower beta-endorphin concentrations than abstainers right after the public-speaking stress. Although Al'Absi's study was not looking at grieving, when we think about pangs of grief, a drop in beta-endorphin levels following a stressful wave of grief could lead to a reduction in dopamine levels in the nucleus accumbens.

This might make smokers more vulnerable to the rewarding effects of nicotine.

The way we feel at any given moment includes physiological aspects of our experience. If we do not pause to consider our next action, as Rizzetto suggests, and reflect on the fact that an urge may be even greater in the moments we are feeling grief or stress, we may find that we are smoking or drinking without realizing it. Without awareness, the desire to shift our state may overcome our best intentions not to cope by using substances.

We Used to Feel So Close

Relationships also have a physiological aspect. Our body makes opioids in response to our long-term loved one, the person with whom we share an enduring bond. This opioid production in a long-term relationship is different from the initial rush of hormones that comes with a new relationship. This may be part of why nothing feels like the relationship we had with a deceased loved one. Obviously, there are many reasons that no other relationship would feel the same, but one of them is that any new relationship we develop is initially using different molecular pathways than our comfy, well-worn relationship. Nothing can replace the relationship history with a sibling or a parent, who has known us our whole lives. A new puppy is different from the bond with one's four-legged best friend, even though they may provide a different sort of joy. Natural opioids reward us for reuniting with bonded loved ones, represent the closeness we feel, and provide that comforting feeling. Part of how we know that long-term bonds specifically utilize natural opioids comes from a study of what happens when we block them with a drug called naltrexone.[10] Psychologists Tristen Inagaki and Naomi Eisenberger conducted a study in which the participants took either naltrexone or a placebo for four days, and then later they took the other one. Participants did not know which one they were taking, and some people got the drug first while others got the placebo first. Importantly, this means that in this study each person served as their own control, called a within-person research design. Therefore, we can compare the participants' reactions when their opioids were blocked to their own natural state, and feel confident that the results arise only from the effect of the drug and not other individual differences we see between people.

During the four days they were taking the drugs, participants kept a daily diary of how socially connected they felt and their general experience of pleasant feelings. Although they did not feel less positive overall during those four days, when they were taking naltrexone they rated their feelings of closeness with others as lower. In addition, the researchers asked participants to come into the lab on the fourth day and do a task to specifically measure their reaction to messages from six people they were close with. Each participant had nominated six people at the beginning of the study (parents, grandparents, siblings, significant others, and friends), and researchers reached out to see if they would write loving messages for the participant. The participant's network was also asked to keep it a secret, so the messages would be a surprise, and the researchers would be able to capture the participant's genuine emotional response. Messages included: "You have enriched my days and given me great joy," "Thank you for loving me at my worst,"

"You're my number one." These notes are so touching; even reading them myself makes me feel warm!

When the participants were receiving naltrexone, ratings of closeness after reading the messages were lower, although their ratings of positive feelings in general were not different. This distinction between just feeling good and feeling close to others is important, and was an elegant part of this study. The brain opioid theory of social attachment suggests that blocking opioids should specifically affect feelings of social connection and not other feelings. This is exactly what the study found to be true. Opioids are about our relationships, providing us information about the connection we have with others and rewarding us for maintaining those long-term bonds. Through the slow process of restoring a meaningful life for ourselves, we should consider that new relationships will feel different from the one we had with our deceased loved one. That does not mean the relationships we begin or strengthen are less important, but it will likely take a couple of years to build up the kind of relationships that will offer similarly warm and comforting feelings our body felt when we were with our one-and-only.

When I consider the current opioid epidemic, I think about how drugs are processed differently in our brain and body when we are grieving. Those suffering from opioid addiction are not necessarily seeking a high, but rather, drugs may resolve feelings of loneliness and disconnection.[11] Given the high rates of grief we find in our culture currently, and the low levels of grief literacy around what to expect during grieving and how best to navigate it, perhaps this partly helps to explain the increase in opioid addiction. If our drug of choice destroys our liver or lungs, this form of coping with loss can be the outcome of a grieving body.

Your Love Is My Pain Reliever

The feeling of closeness affects our experience of pain even when we are apart from our living loved ones. Simply imagining them during their absence has an impact on our physiology.[12] In a study done at the University of Arizona, Kyle Bourassa and David Sbarra asked participants to engage in a *cold pressor test* while their romantic partner was present or while they simply imagined their partner. The cold pressor test involves sticking one's foot into a bucket of ice-cold water—a somewhat painful experience. Surprisingly, both the group who had their partner present and the group who only imagined their partner had significantly lower blood pressure compared to a control group who just thought about their day. In fact, the blood pressure was reduced by the same amount whether the partner was physically present or imagined. Mentally calling on our internal relationship with a beloved person can change our cardiovascular response when faced with a physical challenge like pain.

Evidence of pain relief when we imagine our living loved ones comes from another study, this one by Jarred Younger and colleagues at Stanford University.[13] This experiment used heat at the base of the thumb as the pain challenge, and this was done during a neuroimaging scan so that brain activation could simultaneously be recorded. Each participant had rated their pain response to increasing heat in a visit prior to the fMRI scan, so that researchers knew exactly what heat level each person was willing to tolerate as a brief pain discomfort. Since distraction is also known to reduce pain, in one condition of the study participants were offered a puzzle to take their mind off the pain, and in another condition they imagined their loved one. Both distraction and calling on the mental image of their loved one reduced their subjective reports of feeling pain, but the pain relief worked through different systems of the brain. Only imagining one's partner increased activity in the reward network, and the amount of pain relief was correlated with the increase in nucleus accumbens activity.

What If I Never Stop Crying?

One of the most obvious physiological outcomes of feeling the pain of grief is crying. Surprisingly, science knows very little about why we cry and what the impact of crying is on our body. Even more surprising, there are no studies about crying in a bereaved sample of people (yet). Given how common crying is during grieving, I find this astounding. It strikes me how much we, in our culture, struggle to sit with people who are crying, and this spills over to researchers avoiding sitting with people crying as well. In fact, when my graduate students first start doing studies in my lab, they are often shocked by how much grieving people cry. I have to reassure them that this is normal, and expected, and it will not hurt them, and does not suggest there is something profoundly wrong. For the vast majority of us, crying spells will decrease in frequency and intensity on their own, over time. Crying is the natural expression of grief, on its own timetable.

Thus, given our cultural nervousness about crying, I can understand why people are so reticent to allow themselves to cry, why they hold it in. For the most part, crying is such a strong response that we do not have much control over it. But some people are afraid that if they allow themselves to start crying, they will never stop. Or that they will not be able to bear the pain. But I can tell you this will not happen. Crying, for the most part, actually makes us feel better, or at least calmer, afterward. This is likely because some of those same pain-relieving neurochemicals we have been discussing are released when we cry.

The few studies of crying that have been conducted have discovered there is a difference between irritant reflex tears (like while cutting an onion) and psychoemotional tears (like during a sad movie).[14] Psychoemotional tears contain more prolactin (a hormone associated with nursing and attachment) and opioids (in the form of leucine-enkephalin). Many theories about why we cry have highlighted that it is a form of social communication, that crying visibly allows others to know we need comforting. Although I think this function is relevant, I have a hypothesis that there is another function for crying as well.

Prolactin is stimulated by oxytocin, and we have discussed how oxytocin plays an important role in the grieving body. One role of oxytocin is in attachment behaviors, like nursing and sex. In those two behaviors, part of the role of oxytocin and prolactin is to develop bonds, to learn that this infant or mate is our one-and-only, that we should care for them and remember them and return to them. I hypothesize that crying is another attachment behavior. We know from neuroscience that oxytocin has a *permissive* effect on our brain. This means that while oxytocin is flooding our brain, the likelihood of making connections between neurons increases. Those literal connections between neurons are also how we learn new information. So the presence of oxytocin means we are more likely to learn from the situation we find ourselves in. During nursing or intimacy, this clearly would allow us to learn our loved one's detailed characteristics, in order to remember them and feel motivated to reach out for them. When we grieve and are reminded of the absence of this one-and-only and when we cry, I believe we may also be learning, but learning that they are no longer with us in the physical world. The crying enables our brain to make the connection that their absence is permanent, and although this is extremely painful, it is a part of the process of accepting our new reality and restoring meaning in our life.

Not everyone cries after loss, even when they would like to, and I believe this can be normal as well. Human beings are extremely complex, and we have multiple ways of learning about the world and our changing relationships within it. But if you do cry, every day, even multiple times a day, this is probably just a part of your body's response that is helping it to learn what is happening and find a way through it. Crying deeply and often is not going to hurt you, and in fact, with the deep breathing that often accompanies the end of crying, it may have the effect of helping your body to regulate your heart rate as well.

Many Ways of Avoiding

Drinking alcohol or smoking are coping strategies that allow us to avoid painful feelings. Although these have obvious long-term consequences for our health, other coping behaviors also affect our body. It is not a big stretch to see video games or online shopping as a potentially avoidant coping strategy. These behaviors keep us rooted in place for hours, a coping strategy that affects our body through a hunched-over back or a decline in our overall activity. Binging on junk food is a coping strategy with clear effects on weight and blood sugar. Many of us have known the bereaved person who buries themself in work after their loss, despite their apparent exhaustion. Or the person who is obsessive about their running schedule, even when they are limping around because of damage to their ligaments. Any behavior that pushes our feelings and thoughts of grief out of our awareness can be a coping strategy that will have consequences for our body, because pushing any part of ourselves out of awareness likely means we are also not aware of the impact on our physical self. As I said before, this does not mean that these coping strategies are "bad," but rather, they should be used in moderation or, at the least, we should note the effect they are having on the rest of our body.

Seeing coping behaviors (like overwork or online shopping) as consequential for our body and our health is not a big leap from seeing drugs as a coping behavior with physiological consequences. But whether coping by using particular thought patterns (like rumination) is consequential for our health requires a bit more explanation. In my own life, it took me a long time to learn that how we talk to ourselves and others affects the way we feel, and thus our physical health. It seems so unlikely that something as seemingly harmless as words could affect our health, but like ripples in a pond, the effect of words (in thinking and in talking) reverberates in our body as well as our mind. Words have the power to distance us from our body and our emotions.

I have always loved talking. I talked incessantly as a child. I have a vivid memory of elementary school, my pencil covering blue lines of wide-ruled paper with "I will not talk in class." My hand cramped from the supposedly deterrent punishment of one hundred penalties, but even then I remember thinking, *But that's just not true!* In high school, many hours of chatting happened during the long Montana bus trips for debate meets, at schools hundreds of miles away. I remember, on one of these trips, an introverted friend of mine looked across the seat at me, annoyed, and said, "You talk all the time, but you're just not saying anything." I was so confused by this statement (and hurt as well). To me, talking was how friendship was constituted, how I conveyed my enthusiastic feelings for a friend. With the hindsight of adulthood and a clinical psychology degree, I know now that words were the way I checked in on how everyone around me was feeling. Conversation was my detector, making sure that everyone was okay, that they were okay with me, that I did not need to fix anything. I have to imagine that arising from this subconscious goal, my friends sometimes felt under attack by my barrage of words.

When I arrived at college, I started journaling, and this became another way to express my barrage of words, along with letters to my best friend. Journaling can be a wonderful way to process one's thoughts and feelings, and many grieving people find journaling helpful. Although I did not recognize it then, journaling for me was a form of brooding. I wrote and rewrote about how I felt, why that might be, and my fears that I would never feel better, amplifying my negative, ruminative self-talk. This passive, self-focused rehashing is not the same as processing, which enables a person to *discover* what they are feeling anew in the moment through writing, the cause and effect of changes in their experience over time. Directed writing prompts, like the ones I described in the grief intervention study in chapter 1, are good examples of processing. These prompts can take us in new directions, developing new ways of viewing our experience.

From this history of incessant talking, you can imagine how shocking the experience of a silent Quaker meeting was for me. When I refused to get confirmed as a teenager, my mother wisely told me that I still needed to have religious community, even if it was not Catholic. My best friend was Quaker, and when I asked her what they did, she told me that a small worship group sat together in her parents' living room in silence. If no one was going to tell me what I had to say or believe, that felt like a religion I could live with. So each week I sat in silence with her family and ten or so other adults for an hour, in an old Victorian house with floor-to-ceiling bookcases. It was profound, and confusing, and sometimes vaguely upsetting—I occasionally found myself tearing up for no apparent reason, which did not seem to surprise or bother anyone. When I finally asked someone what she was doing while sitting in silence, she told me she was listening. As a teenager, I had no idea what that meant, but I tried to listen, not really knowing what I was listening for, or where the listening should be directed.

Sitting in silence as a teenager was the beginning of a long journey of discovering how my mind works and how to achieve more balance in my state of mind, more peace and calm. I discovered that using words can be a way to avoid feelings that might otherwise well up, because my tears welled up when I stopped talking and just sat. It was my first inkling that there was a state of just being, rather than always doing. When I developed significant fatigue because of MS, I no longer had the energy for so much talking, so much social interaction. This meant I had to face whatever feelings arose if I was not running around, running my mouth, thoughts running through my head.

Forced to slow down, I recoiled at another idea that crept into my consciousness: Was there something I was avoiding by using all this verbalization? What was I so afraid of?

Thoughts Can Affect Our Body

In the decades since I first realized there was a connection between what I was thinking and my physical state, I have studied the research on rumination. It may surprise you that I can now list several different mechanisms through which words affect our physical state. First, there is the *content* of words. Negative self-talk (e.g., *I'm such a loser; I'm never going to find someone who loves me*) affects our emotional state, and our emotions always include a physiological component. We feel anxiety in our body, and our body enables the feeling of arousal or tension. We feel depression as lethargy in our limbs, preventing us from pursing meaningful activities. This first mechanism through which words affect our physical state is the lever used by cognitive behavioral therapy (CBT). CBT can improve our physical health as well as our mental health. This psychotherapy method teaches clients to become aware of their automatic thoughts, and then the effect of their automatic thoughts. Together, therapist and client evaluate whether a thought is true by examining evidence for and against the thought (e.g., *Everyone goes through times when they are not in a relationship, so maybe it's not because I'm a loser*), which consequently enables the client to change their self-talk, often changing the way they feel both emotionally and physically.

A second mechanism through which words affect our physical state is *maintaining* perseverative thoughts. For example, if you bump into a colleague, and they offer their condolences on your loss and then say, "Well, at least they are in a better place," it might make you angry. But if you continue to run that conversation through your mind, over and over, even after the conversation has ended, your blood pressure and heart rate will likely remain slightly elevated. In this way, your body continues to confront a situation that is no longer happening in the present moment but *is* happening in the virtual world you are inhabiting through your thoughts—the thing they said, the thing you wish you had said. The words can hold you (and your physiology) hostage to the past, even if the past was just a few hours ago.

A third mechanism is the brooding nature of rumination. Rumination goes on and on, the litany of thoughts about our grief or about our loved one's death, often with a notably self-focused perspective. *Why do I still feel this way? What is wrong with me? Is this normal?* You may also notice that these thoughts are passive—rumination is not a problem-solving exercise but rather endless reviewing without any plan to cope differently in the future. In *The Grieving Brain*, I mentioned that the would've/should've/could've thoughts that often occur after a loved one dies is a common experience of grief-related rumination. Perseverative thoughts maintain our reaction to an experience that just happened, but rumination can occur at any time, especially when nothing else is going on (like sitting at a stoplight or when we get into bed at night). Just discovering what rumination as a mental state feels like, and remembering to check in with our mind to see if we are ruminating, is half the battle to learning to let go of these thoughts. Rumination is remarkably unremarkable, and it can be hard to notice when we have slipped into this spiraling thought process.

Rumination as a Form of Avoidance

One way to understand the consequences of rumination is to consider these key questions:

What am I not *doing because my attention is caught up in endless rumination?*

What would I be experiencing if I were not stuck in my thoughts?

A possibility is that we use brooding as a (subconscious) way to keep something out of our awareness, such as fearing an uncontrollable wave of grief. Because grief is so painful, the thoughts racing around in our mind might distract us from any feelings that may arise, keeping us in *thinking* mode and out of *feeling* mode.

In fact, grief psychologists Maarten Eisma, Maggie Stroebe, and Henk Schut at Utrecht University formalized this possibility into a testable hypothesis in the 2010s, which they call the *rumination as avoidance hypothesis*.[15] They pointed out that during rumination our mind may be using our internal dialogue to avoid whatever else it would be thinking (and, more likely, feeling). Any good hypothesis is followed up with experimental testing. But the challenge with measuring avoidance is that the exact process one is trying to measure is outside of the participant's awareness, and thus not accessible to the self-reporting we often use in psychology studies.

Eisma and colleagues found an ingenious way to measure subconscious awareness. They used special headgear to track the movement of bereaved participants' pupils as they looked at side-by-side photos of their deceased loved one and a stranger. The amount of time spent looking away from the photo of the deceased was used as a quantifiable measure of avoidance. Since a shift in eye gaze happens on the order of milliseconds, it is outside of conscious awareness, and thus, it taps the unconscious process the researchers were seeking to measure. Sure enough, bereaved people who reported ruminating more also spent more time looking away from the photo of their deceased loved one, demonstrating that rumination and avoidance are linked.[16]

These study results still do not explain exactly *what* we are trying to avoid, to keep out of awareness, through the process of rumination. Likely what we are avoiding is unique to each person. Grief means there are some likely candidates—a brush with death, rejection, or a loss probably initiate thoughts and feelings we do not want to allow entirely into our awareness. The realization that we ourselves are going to die, that no one lives forever, is another candidate. The realization that our other loved ones will someday die, that no one is immune from mortality, even our children. The realization that our health is on borrowed time, and we will inevitably experience loss of capacity and energy. The realization that we will feel alone—deeply, existentially alone—at times. Just putting those sentences down in black-and-white makes me consider avoiding writing more, simply because they sound so overwhelming. But I can tell you, I persist now in writing the rest of this book because my experience is that neither protesting the awful truth of these realizations nor despairing because they are true is a way forward. Stuck in protest or despair is no way to live one's life. Protesting the truth of these realities only leads our body to react to the anxiety they produce without the benefit of our conscious mind to help us cope. Despairing that these truths mean the future is hopeless denies our human capacity for healing, growth, and learning to thrive in the face of them, and to act with more compassion and love for our fellow human travelers who must also live under these realities.

When my mother was terminally ill and I was overwrought, I sought out a counselor. I remember saying to my therapist, uncharacteristically, "Well, she's going to die. I'm not sure there is much else to say." I was feeling a lot more than I could put into words. I had the sense that a lot was happening inside of me, but I could not find the words; I felt unable to communicate my internal experience. Of course, the irony is that I am still talking and writing about the grieving process more than twenty-five years later. Even if our culture does not encourage it, the nature of being human means you may want to contemplate these awful truths as I did, these painful realities we all must live with, and try to find a way to describe how you feel about them. And through this process, you may find a way to accept the realities and move forward with your life.

Grieving Can Be an Opportunity for Healing

We may have been using avoidance for a long time, long before our loved one died, and it may have been affecting our health to some degree. But bereavement heightens the consequences of using avoidance, because the experience of loss and grief that we must accommodate in our new reality is so intense and disruptive. The effect of not learning to accommodate and accept our loss increases our risk of illness, disease, and even death. Avoidance has many faces: It can look like drowning ourselves in a bottle, smoking like a chimney, working beyond our body's physical limits to the point of pure mindless exhaustion, or brooding on the same would've/should've/could've thoughts over and over, without figuring out how to cope with the painful truth of the loss in the here and now. But grieving is an opportunity for healing. By sitting quietly and listening, we may be able to discover how we feel, what we need, what we believe, and what we value, as well as how to build a healthy new life that incorporates those deep and true motivations.

CHAPTER 5

The Brain

When I talk with a group of bereaved people, someone inevitably mentions brain fog. I can understand this—it is incredibly frustrating to have your mental capacities clouded. You also may feel like you are losing your mind, since brain fog is not usually named among normal reactions to the loss of a loved one. And yet it is very common and has even been the subject of bereavement research in recent years. But the first thing to know is that not everyone means the same thing when they use the term *brain fog*. In fact, when grief researchers study it, they are often using neuropsychological tests of memory or processing speed, or other discrete mental functions. These are objective measures of cognitive ability, which they can then compare between bereaved and nonbereaved people, or look for a trajectory of change over time in a person.

Unlike a single measurement of cortisol or inflammatory markers, neuropsychological tests are sensitive to small changes and are benchmarked. This means an individual's score can be compared to what is expected for a person's age and education. But even if a person's objective score on a mental test is normal, neuropsychologists recognize that the subjective experience of feeling mentally foggy may still be real. So here I will unpack the range of dimensions and some causes of brain fog, while acknowledging that this is still an active topic of research and clear conclusions have yet to be agreed upon among grief researchers.

Demonstrating that different people mean different things when they say brain fog, Laura McWhirter and colleagues at the University of Edinburgh did an innovative study. Posts from the social media platform Reddit were monitored for the term for seven days in 2021, to capture the way people spontaneously self-describe the experience.[1] They discovered that the term is used to describe a range of phenomena, including forgetfulness, difficulty concentrating, dissociation, cognitive slowness and excessive effort, communication difficulties, "fuzziness" or pressure, and fatigue. These posts were in subreddits covering a wide range of illnesses (including COVID and psychiatric, neurodevelopmental, autoimmune, and functional disorders), drug use and discontinuation, and other situations. This makes sense, since brain fog happens in a variety of circumstances, including bereavement, multiple sclerosis and lupus, pregnancy and menopause, chemotherapy, insomnia, chronic fatigue syndrome, and depression. Most recently, brain fog has been described as a symptom in long COVID, one of the debilitating symptoms that may develop following a coronavirus infection.[2] From this list, you can see that when people use this term, it does not tell us much about exactly how their mental state is being affected, let alone why they might be experiencing it.

What is clear is that brain fog is not "all in your head," literally or figuratively. Brain fog, like physical pain, is a subjective experience, and even when medicine and psychology can identify brain and body markers associated with it, it remains a phenomenon that requires self-reporting, although we can measure it. The quantitative assessment of a subjective report (how much brain fog are you feeling on a scale of 1 to 10?) waxes and wanes reliably, and quantitative self-reporting reliably predicts other outcomes. All this is a fancy way of saying brain fog is real. The fact that science may not yet have discerned all the causes or treatments makes it no less valid.

Nonetheless, if you are experiencing brain fog during your grief, knowing the range of possible causes may make it easier to adjust (or accept) contributing factors. For example, the dull roar of insomnia plagues many during bereavement and may make it hard to concentrate or make decisions. When your schedule has changed completely from caring for an ailing parent or beloved pet every hour, the push-notifications your brain continues to erroneously send you may interfere with what you need to remember in your new reality. Drifting off into what-could-have-been can affect any ongoing task, like your ability to concentrate on paying the bills. Now is the time to use a daily pill organizer, Post-it notes, a phone alarm, a calendar, a to-do list. You might take a photo on your phone of where you park your car or ask your teenager to text you where they are going after school, even if they already told you this morning. This does not mean you will have to use these compensatory tools forever, but bereavement is a time to use all the supports you can.

Given that this chapter is all about the brain, however, I want to fill you in on some exciting new discoveries that neuroscience is making about *how* and *why* brain fog may occur during bereavement. The brain is an organ, like any other organ in our body. Consequently, the brain is subject to the same impact that inflammation or hormonal changes can have on any organ. It has architecture that can be remodeled, and its electrical signals can be slowed or misdirected. The brain can be studied by measuring its electrical waves and blood flow patterns, and its physical structure. Any of these assessments can show changes from normal structure and function—but notice that these measurements are still in a different category from thoughts, feelings, and behavior. Phenomenological experience is a different domain that may or may not directly map onto the physical changes seen in the brain. Science continues to discover the connections between the brain's hardware and the reality we each experience. But our experience remains inherently subjective, emerging from the physical gray matter, and experience will always be a different thing from physiology. Even brain lesions in MS, white on black in an MRI, may or may not show up as symptoms in the person's functioning. Assessing both the physiology and the psychology will always be important—one will never be able to stand in for the other.

The Brain Is Not Immune to the Immune System

I wrote an entire book about how the brain is involved in grief and grieving—*The Grieving Brain*—so it may surprise you that there is more to be said on this topic. But in that volume, I primarily discussed the impact of grieving on the mind, how the brain encodes our bonded relationship and then must update its internal map of the world after the loved one has died. I offered insight into how we can be torn between two streams of information at the same time—the memory of the death and also the attachment belief that our loved one will always be there, the *gone-but-also-everlasting* theory. I proposed that grieving is a form of learning, learning that takes a very long time. In contrast, the chapter you are reading now considers the brain as just another organ of the body, and the impact of bereavement on neurophysiology that can affect us beyond what is in our mind's conscious awareness.

For decades, medical science believed the brain was an *immune-privileged* organ, untouched by the comings and goings of our immune cells because of the blood-brain barrier. Work done in the field of psychoneuroimmunology in the 1990s and beyond revolutionized our understanding, finding that the immune system has multiple roles in the brain. This is important because if we are not even looking for the impact of inflammation on the brain, for example, we are not going to discover whether it plays a role!

First, there are specialized immune cells for the brain called microglia. I like to think of microglia as the resident gardeners there. Much like a gardener notices when a windstorm has damaged trees and bushes, and may trim them so they will grow back with a good structure, microglia cells react to injury (like stroke or traumatic brain injury) by creating scars, removing dead cells, and encouraging new neurons to develop and compensate for lost functions. In this way, our microglia function in the brain the same way other types of immune cells do in our other organs. Microglia cells produce inflammatory markers as well, like a gardener sending radio messages requesting more help in a particular area. In the next section, we will discuss what we know about the role of microglia in bereavement from animal studies and imaging studies of the human brain.

Second, we now know that our body's peripheral immune cells—in addition to resident immune cells in the brain—get through the blood-brain barrier in certain places and under certain conditions. Although this can be useful if there is an infection in the brain, it can also be problematic. In MS, the blood-brain barrier becomes permeable following immune activation in the body. This disruption allows immune cells to gain entry into the brain, akin to having additional overzealous arborists show up in the garden. These nonresident immune cells produce inflammation in the brain, causing damage to neurons and the cells that support them, which results in MS symptoms and changes to the person's functioning. You can imagine how peripheral immune activation during acute bereavement could set off a whole domino effect if a person has risk factors, resulting in MS-related impacts in the years that follow a loss, perhaps like my own.

The third role of the immune system in your brain is perhaps the most fascinating. Our brain is actually in charge of leading and coordinating immune system functioning throughout the whole body. In that role, the brain needs ongoing information about the current situation in all parts of the body. This means that information about inflammation produced in all parts of our body is communicated to our brain, even without the entry of actual immune cells across the blood-brain barrier. This information can be communicated in several ways. Inflammatory proteins produced in the body, cytokines like IL-6, can directly cross the blood-brain barrier in certain places. But additionally, inflammatory proteins bind to receptors on our vagus nerve (which you will recall "wanders" through every organ of the body). The vagus then transmits electrical signals to the brain, relaying the information about inflammation levels in specific organs and body regions. Our remarkable brain takes in all of these forms of chemical and electrical information and makes assessments of the current bodily state. In addition, because our brain is a prediction organ, it uses information about our social and physical environment to anticipate what immune responses might be needed in the near future. All of this usually results in an amazingly smooth and effective brain-immune coordination for the body—but you can imagine there are many ways that things could go awry.

The Impact of Peripheral Inflammation on the Brain

My own psychoneuroimmunology studies at UCLA investigated the role of the immune system in bereavement, and I discovered that inflammation is increased in bereaved people, on average.[3] In a second analysis that came from that same study, psychiatrist Michael Irwin and I assessed whether there was an association between inflammation in the body and activity in specific brain areas. Although inflammatory markers are usually measured from the blood, I was curious as to whether we could use saliva to measure them in the mouth, since sampling saliva is much easier for participants than a blood draw. Saliva is also a bodily fluid relevant to our health—among the many other impacts of bereavement, widows have higher levels of periodontal disease.[4] When we measured saliva in our bereaved participants, we found greater inflammation in the mouth than a nonbereaved group. For some bereaved participants, the inflammatory markers were in the range we would expect for periodontal disease.

These same bereaved participants had also undergone neuroimaging scans while looking at photos of their deceased loved one and grief-related words.[5] In a comparison condition, they looked at photos of a stranger and neutral words. Remarkably, we found a positive correlation between the level of inflammatory markers in saliva and the amount of brain activation in a specific region. This region is called the subgenual anterior cingulate cortex (sgACC), a brain region regulating mood. Note that this result does not mean there was inflammation in this region of the brain. This finding means *information* about the level of inflammation in the mouth was communicated to the brain, and those with higher inflammation in the mouth showed more activity in a part of the brain controlling mood.

After the scan, we asked the participants if they could recall any of the words they saw while they were in the scanner. Those with higher sgACC activation were able to remember more grief-related words compared to those with less activation in this area. To ensure they did not just have a better memory, we also asked if they could remember any neutral words. The number of neutral words they remembered was not related to their level of sgACC activity. This means they showed a bias toward remembering grief words. One function of the sgACC is to monitor the emotional meaning of environmental cues. We interpreted the association between greater sgACC brain activation and better memory for grief-related words as evidence that those with the highest inflammation are motivated to preferentially attend to grief cues in their environment. Once again, we see that the health of the body influences what we pay attention to and how we feel, even if we are not aware of it.

Inflammation in the mouth is our healthy immune response to bacteria, a normal response to manage the germs that we regularly encounter. Periodontal researchers have pointed out that while microorganisms can damage tissue in the mouth and trigger periodontal disease, germs alone are not sufficient for periodontal disease progression.[6] The experience of stressful life events like bereavement may play an important role in susceptibility to infection, such as the kind leading to periodontal disease, when the immune system cannot protect us sufficiently.

In response to bacterial infection, your immune system increases inflammation, in a good and necessary way to defend and rebuild. Your immune system enables healing through this communication of inflammatory signals. However, the slow process of healing, and the inflammation that it requires, influences our emotions, thoughts, memories, and behavior. What science is beginning to understand is that this healing process applies to stressful psychological events like loss. Bereavement leads to immune activation, which requires time and healing, just like for any other physical illness.

Until recently, we knew little about how the combined immune and brain processes contributed to grief or depression. New knowledge, however, opens doors to new interventions. Researchers are discovering that antidepressants may have part of their effect through reducing inflammation, and that anti-inflammatory medications can improve our mood. In fact, we found when we gave aspirin to acutely bereaved people to help their heart health, they also reported slightly lower grief, compared to those taking a placebo. The psychic pain we experience in bereavement may be a part of the process of healing after the amputation of our loved one. With future research, we may come to better understand how to support the immune system during this healing process.

Today's Brain Fog

The enormous irony is that today I am trying to write this chapter while I am experiencing brain fog. As with most brain fog experiences, this is probably due to multiple causes. I recently received a vaccine booster, which often results in briefly higher inflammation and can temporarily increase my MS symptoms, including feeling foggy. I also have been having hot flashes, problematic because they keep me from sleeping for more than a few hours at a time. Sometimes having an extra cup of coffee is sufficient to overcome the fog, but caffeine also leaves me feeling restless, and my mind, already stretched by normal thinking during brain fog, tends toward anxious thoughts that jump around and make concentration difficult. In addition, in Tucson, Arizona, where I live, we are currently under an excessive heat warning, with ten days in a row over 100°F, and heat slows nerve conduction for those with MS. The brain fog makes it hard for me to carry a train of thought, often having to stop and start, constantly needing to remind myself what I need to hold in my working memory for a task. Many different conditions contribute to brain fog in bereavement as well.

Sometimes during these episodes, I recognize that I am simply not going to get through a statistical analysis or write a logical paragraph, and I give up altogether. Today, I am dictating this chapter instead of typing (and note that what I write here will be edited later, on a day with better mental faculties!).

So with years of experience with occasional brain fog, what do I do? Most important, I remind myself that this mental state will not last. This is the case with grief as well. A hallmark characteristic of waves of grief is that they are exactly that: waves that come and go. But sometimes I start worrying that today's brain fog means I am doomed to a future of cognitive impairment. This worries me surprisingly often, given that I have managed MS for a quarter of a century. When this happens, I turn to accepting my current limitations as a coping strategy: *Sometimes I have brain fog.* If this was my mental state for the rest of my life, I have come to realize that my life would still be pretty wonderful. Brain fog does not keep me from watching the hummingbirds outside my window, enjoying a piece of delicious dark chocolate, or stroking a purring cat. Even with brain fog, I can still hike in my favorite mountain canyon. I do have to be more conscientious without all my mental faculties— did I bring my water bottle, ID, energy bar, keys? I have to use my compensatory strategies and habit serves as one resource. I have done this hike many times before; I know exactly where I will park and which trailhead I will enter.

Having worked in hospice transforms one's idea of what makes up a good day, and a good quality of life. If I was no longer able to work as a professor or an author, my grief would be overwhelming. But I also know that I would eventually find a way to thrive, even in the absence of my life of the mind.

Brain Fog Will Not Last

Because I can remind myself that these few days of brain fog will not last, I also try to keep my schedule as routine as possible. I drag myself out of bed at the same time, eat the healthy breakfast that I eat every day, have my cup of coffee, and take my many medications. I shower and go to work, even if when I get to work, the first thing I do is to take a nap. Habit truly matters, and I am trying to remind my brain that this is the daily structure it can return to when it is ready.

But I also recognize that this is not the same as my regular day. Most importantly, I have to find a way to reduce my expectations of myself, the demands of my world. This means checking my calendar for the one or maybe two things that absolutely must happen today, and getting those done first, or asking for help with them. And then I give myself permission not to think too hard. I respond to my sadness about this brain I have been entrusted with for a lifetime (for better and for worse) by trying to comfort myself, as I would comfort a dear friend. You might respond to your sadness over the death of a loved one this way as well. I try to be gentle with myself, offering and allowing myself to enjoy simple things in the day.

This is quite in contrast to my younger self, who had a way of trying to punish myself for being lazy, or feeling stupid, by either battling with mental tasks I could not complete or forcing myself to do monotonous chores. My theory was that if I got those chores done while I was mentally fatigued, then I could work even harder, or for longer, when my mental faculties returned. I will explain more about how I got to this place of greater acceptance, greater equanimity, in the last four chapters, but suffice it to say I have learned that punishing oneself does not help one's mood or mental state.

So I rely on the checklists and schedule that I consistently plan in advance, a resource I offer my future self because I never know when I might wake up on a day like today and not be able to think straight. But I move those tasks to other days (or even the afternoon), and take the time to go and buy flowers or take a walk or sit in quiet comfort. I try to notice what my body seems to want to do, like following the lead of a small child who is playing. Like watching the rise and fall of my breath when I meditate, there is much wisdom in my own body and subconscious mind, and I try to allow them to be in charge and help my brain to heal.

You, too, may come to recognize that being bereaved means accepting that you will not know on which days that grief will overwhelm you. Your grief may come with brain fog or fatigue or unbearable sadness, and you may need to build habits that will support you during this time, combined with gentle self-compassion for simply getting through the day.

Microglia: The Brain's Resident Gardeners

One cause of brain fog is higher levels of inflammation in the brain. As I mentioned, microglia are the resident immune cells within your brain. Like gardeners, they provide the regular pruning and weeding needed for normal functioning, in addition to fighting infections and responding to neuronal damage. Microglia were first described by Spanish neuroscientist Pío del Río Hortega in 1919, and they make up 10 percent of the total cells in your brain. Their role includes ensuring the survival, repair, and remodeling of neurons. When there is damage within your brain, like from a stroke or head injury, microglia migrate to the area of damage and release inflammatory markers, facilitating the influx of immune cells from your body if needed. Unfortunately, although the influx of immune cells is helpful, it can also have negative effects, even causing difficult cognitive and emotional changes.[7] In addition to traumatic brain injury, microglia are responsive to both acute and chronic psychological stress.[8] As these microglia cells respond to the world around them, they change shape and size, like gardeners putting on overalls and making their pockets bulky with extra tools when they arrive at work. This change in their shape and size enables researchers to see whether they are activated in response to an immune event, in contrast to their smaller size when they are quiescent. In neuroimaging studies of people who are depressed (although not necessarily bereaved), scientists can see activated microglia cells in the same sgACC region of the brain are correlated with depression severity.

Notably, the neuroinflammation created by microglia interacts with the production of neurotransmitters like serotonin. For example, some antidepressants reduce the activation of microglia cells and thus reduce neuroinflammation. In chapter 2, I talked about the molecular pathway that enables our body to create serotonin from the tryptophan protein we eat. But under conditions of chronic stress, instead of making serotonin from tryptophan, another toxic byproduct—called quinolinic acid—is sometimes made instead. Why would the body make a toxic byproduct? Ordinarily, quinolinic acid is needed for the normal functioning of microglia to prune synaptic connections, like a gardener might use a small amount of herbicide.

But dysregulated pruning might occur under chronic stress. Synaptic over-pruning can lead to weak transmission of nerve signals, decreased functional brain connectivity, and deficits in social interaction.[9] The toxic byproduct produced by microglia is found in higher concentrations in the sgACC brain region of those who died by suicide, seen after autopsy.[10] Certainly not all bereaved people experience depression, but bereavement does increase the risk of depression. Perhaps one day we will discover that microglia are one link in the chain between a stressful loss event, immune activation through microglia, increased neuroinflammation, changes in serotonin levels, and depressive symptoms. Bereavement may also be a time when rewiring of the brain, and thus synaptic pruning, is needed for adaptation to life changes. But this normal process might become problematic in some situations, and I hope grief researchers will take up this line of reasoning in the years to come.

The Gut–Brain Axis

Pharmacological antidepressants are not the only compounds that have an impact on depressive symptoms. Some plant-derived natural compounds affect depressive behavior (at least in animals).[11] These include flavonoids (found in vegetables) and terpenoids (found in fruits and vegetables). I find the role of food on mood to be particularly interesting given the changes in appetite and weight so often seen during grieving (described more in chapter 8). Since the 2010s, researchers have been investigating the body's microbiome. As the National Institute of Environmental Health Sciences defines it, our microbiome is: "the collection of all microbes, such as bacteria, fungi, viruses, and their genes, that naturally live on our bodies and inside us. Although microbes are so small that they require a microscope to see them, they contribute in big ways to human health and wellness. They protect us against pathogens, help our immune system develop, and enable us to digest food to produce energy."[12]

In addition to my interest in how inflammation in the mouth affects the brain, other researchers have been interested in how inflammation in the gut affects the brain. Neural pathways from our gut to our brain make a lot of sense, because our gut is one of the places where we are most exposed to the outside world, through the food we consume. Our immune system needs careful surveillance of what is ingested, to distinguish what is nutritious from what is dangerous to us. Since our brain coordinates these immune responses, the information relayed from gut to brain, and brain to gut, is vital for healthy functioning. The gut-brain axis often works through the microbiome. Stress is one of the factors that can cause a disturbance in the balance of gut microbiota. This stimulates gut-to-brain transmission of information about local inflammation and can lead to microglia activation.[13] When there is an imbalance between beneficial and harmful bacteria in the gut, it alters the permeability of the gut barrier, literally creating what is termed a *leaky gut*. This allows bacteria and their byproducts to activate the immune system, leading to inflammation and potentially neuro-inflammation.[14] The microbiota of the gut change in response to external situations, such as diet, use of antibiotics, infections, and stressful life events, like bereavement. Thus, the gut influences brain function, especially brain areas involved in memory, learning, emotion, and mood regulation, leading to feelings of depression, anxiety, or brain fog.

The Gravity of Being Off-Balance

Brain fog does not only affect how we think. When our brain is not working smoothly, it affects all of our thoughts, feelings, and behaviors. This includes unconscious reactions and basic functions, like balance. For example, when Lauren DePino's maternal aunt died, she says it was hard to explain the enormous impact it had on herself and her whole family.[15] Her aunt was the matriarch of their large Italian family. She organized the holidays, called to check in, and was the glue that held everyone together. But it was a year later, near the anniversary of her death, when Lauren fell on the stairs of her patio. Bloodied hand, bruised knees, and the broken mug she had been holding led her to wonder why she had fallen, and to reflect on her continued grief, her dazed state. When she related the spill to her mother and her sisters, she found that they had also taken falls that week. Both our fuzzy mental state and our exhaustion while we are grieving are normal but also can be dangerous. Our balance is both figuratively and literally affected after loss.

Lauren is a writer and a singer, specifically a funeral singer. Having learned to detach from her feelings when she performed at funerals, she realized that she had been doing the same with her own grief over her beloved aunt. She says she had been living in a dazed state. Remarkable to me was Lauren's determination to make this crash into her real world and onto her patio tile an opportunity for healing. A trip to her familial home in Italy followed close on the heels of her fall, and her determination to stay in the present moment, to learn to give up her detached state of coping, was facilitated by the precipitous stairs carved into the rock foundation of Amalfi, the medieval town on the Italian coast that was her family's origin.

Lauren describes how, as she climbed these tiny, winding, unpredictable stairs, she could not have imagined the fierce concentration it would take to keep her from toppling into the sea. She made the connection that she needed to apply the same intentionality to her forward motion in her grieving. She writes, "Grief can creep into our lives, months—even years—after our loved one has died. It can besiege our most joyfully anticipated experiences until we no longer see them as joyful. Not until we pay grief the attention it seeks can we live again."

Lauren points out that becoming disconnected from our body, from our feelings, from the world around us, makes it hard to stay upright while putting one foot in front of the other. Whether our difficulty concentrating comes from neuroimmune processes in our brain or a habit of avoidant coping, we can work toward giving ourselves more time to process the world around us (and grace when we cannot) and toward practicing present-moment awareness.

Lauren is not alone in the impact grief had on her. One particular statistic has always stood out to me in the large Finnish epidemiological study by Pekka Martikainen and Tapani Valkonen that I described in the introduction. This study reported increased mortality risk for widows and widowers, broken down by the cause of death. Fatal motor vehicle accidents were 30 to 40 percent higher for widows and widowers than the married comparison group.[16] In addition, "other fatal accidents" were more than twice as likely for widows or widowers than for married people. This category is distinct from deaths by suicide during bereavement in the study's analysis. I have long wondered if our inability to concentrate, to pay attention, to be aware of our surroundings during grief, leads to this increase, but no research to date has examined this possible link between the cognitive state of grievers and the incidence of accidents.

You Are the Expert on You

Western science is a very powerful tool, a method to cordon off a particular problem and, holding all other things constant, offer an answer that is reliable. But "holding all other things constant" often finds itself at odds with the larger reality. Everything we have questions about—our body, our emotions, our relationships, our health, and beyond—have answers that come from a larger context. So the scientific answer works in the area cordoned off in the empirical studies but may not make sense in the larger dynamic system in which it is embedded.

For example, studies of the broken-heart phenomenon focused on blood pressure and adrenaline have not usually also considered the ways that the person's perception of demands and resources results in the experience of bereavement stress. Or, studies considering the amount of inflammation created during prolonged grief have not included measures of whether inflammation is also the cause of social withdrawal, exacerbating the feelings of grief, and recursively contributing to worse health. Much of what I have been telling you about the history of bereavement science in this book—from early immune studies in the 1970s to attachment theory in the 1980s to psychoneuroimmunology in the 1990s to my own cognitive neuroscience understanding of grief since the turn of the twenty-first century—has been an attempt to place our body, our emotions, our relationships, and our health in a larger context. Our historical era of bereavement science is in response to earlier psychoanalytic and behaviorism ideas focused on the reductionism of systems of the body. The body was treated as separable from the mind. Grief research from our era of history has demonstrated evidence of the mechanisms that tie a person to their beloved other (through powerful neurochemicals that motivate feelings and behavior), tie grief emotions to the body (through ruminative thought, avoidance, and awareness), and tie the gut to the brain (through the immune, endocrine, and peripheral nerve pathways, a topic we will continue to explore in the following chapter).

Each of the answers science gives us provides data only for the questions asked in that particular study. Science can tell us through randomized clinical trials if a psychological intervention changes the bereaved group's grief severity. Science can also tell us whether giving anti-inflammatory medications affects a bereaved group's grief severity. But science cannot necessarily tell us what an individual grieving person needs. The question of how to achieve health, how to heal, is both a larger question for society and a unique question for each person. In many ways, each person needs to discover how to heal, albeit with the benefit of those group-level scientific grief studies. On the other hand, I and a growing number of others believe that approaching grief as a public health issue would be a productive new lens that might prevent prolonged grief for many people.[17]

Most of us experience pain when someone important in our life dies. This is natural, and for most of us, waves of grief decline in intensity and frequency across the first year. At about a year, clinicians can begin to see that there is a very small group (less than 10 percent) who do not show any change over time, and we call this prolonged grief disorder. After a year, their grief looks the same as it did in the first weeks, or even worse. No one stops feeling grief after a year, but at that point, clinicians can see what grieving trajectory people are on. For those who are not showing any change in their waves of grief, research predicts they are not likely to show change in the future either. For those who are showing growing acceptance and adaptation to having grief, even if their life is not restored to the way they would ultimately like it to be, they will likely continue on a trajectory of change.

For people experiencing grief, primary interventions can be accomplished with wide-scale, society-level education. Secondary interventions, like bereavement support groups or mobile health apps, are for bereaved individuals who, through screening or assessment, could be regarded as more vulnerable to the risks of bereavement (e.g., high levels of distress, low social support or isolation, developmentally sensitive periods). Tertiary intervention might focus on precious resources for individual therapy for those suffering from prolonged grief.[18]

I believe that just because we have identified a region in the brain where the loving bond between two people is encoded, and we hypothesize the oxytocin and opioid binding there decreases during bereavement, it does not necessarily mean we should intervene by trying to change the brain directly, as with medication. Rather, it points to the incredible importance of the attachment bond, notable because of the amount of resources the brain devotes to maintaining this bond, conserved even across species. The brain has evolved to protest the absence of the beloved in case they could be recovered, to despair at their absence and what that means for the person's life, and also to adapt, transforming the internal working relationship with the beloved and restoring a life in which love and community are possible again. Perhaps the best course is to support this natural process, discovering and recognizing the ways it can go awry, but intervening by supporting the relationships, thought patterns, and behaviors that improve the bereaved person's experience. In many ways, I believe science can reach the same conclusion that Indigenous communities have known all along through an entirely different system of knowledge: that the people, the community, are the medicine.[19] If we want to increase oxytocin or opioids, why not support the bereaved person's feelings of mattering and belonging, their opportunities for physical touch, which demonstrably lead to their body's own oxytocin and opioid production?[20]

But notice that I said, "I believe." My wisdom as to how to approach grieving may be different from yours. The data from laboratory science or even intervention science tells us very little about you, specifically you. Science offers more information than you could have without those studies—information on how grief works in most people, on average, and on the unconscious processes (including implicit beliefs, habits, and physiological correlates). That might mean grief works differently from how you thought it worked before you learned of the studies' results. But while I may be an expert on grief, you are an expert on you. Scientists, psychologists, medical doctors, yoga instructors, and clergy are only consultants, and you must listen to your own body and mind—and perhaps your relationships with your friends, family, neighborhood, and community—to determine what works for you.

CHAPTER 6

The Sympathetic Nervous System

When my mother received her stage IV cancer diagnosis, had a mastectomy, and began a year of weekly chemotherapy, her friends were at our house more often than I could ever remember. I was in junior high, and it had not really occurred to me that my mother had friends, outside of the parents of my friends. But her close friends turned up to support her in many tangible ways. One of her good friends had been an Olympic skier and had faced her own bout of breast cancer in the previous year. On the advice of her former ski coach, she had gone to the University of California, Berkeley, for cutting-edge medical treatment. This friend of my mom's brought back Herbert Benson's book *The Relaxation Response*, which was being used at academic medical centers to address high blood pressure and chronic pain, and in cancer centers to counteract the stress of cancer treatment. My mother was willing to try anything that might improve her odds of beating metastatic cancer (including giving up red meat and black tea—no small feat for a British woman!). I truly believe that these changes she made, developing new daily habits motivated by the desire to live for her two teenage daughters, extended her life; I remain eternally grateful to her for her superhuman feat. Once a day for twenty minutes, my mother would lie on the recliner in our living room, eyes closed, listening to a guided meditation tape. An awareness of the intertwining of mind and body was only just taking hold in our national psyche in the 1980s, but her experience influenced my thinking for decades to come.

Having witnessed the impressive effects, when I did a rotation at the UCLA/Revlon Breast Clinic as a psychology intern in 2003 and came across a guided meditation CD by Belleruth Naparstek, I began listening to it every day. I developed the ability to adjust my conscious state by listening, by training my awareness on her words, by learning to maintain my concentration there rather than on the myriad other thoughts my mind could generate. I became conditioned to relax whenever I heard her voice, and it still calms me to this day.

A Link Between Mind and Body

The guided meditation my mother practiced, and resulting relaxation, likely reduced her noradrenaline levels. It probably also contributed to the fact that her cancer did not recur for another decade or so, although her surgery and chemotherapy no doubt had a larger impact than the relaxation practice alone. The mechanism for how relaxation might physically help was not understood back then, but recent studies show high levels of noradrenaline can promote the growth and spread of existing cancer.[1] Sympathetic nerves travel throughout our body and into every organ, similar to the wandering vagus nerve. Sympathetic nerves release noradrenaline into organ tissues like breast or liver. (Noradrenaline is also called norepinephrine and you might be more familiar with its closely related cousin, adrenaline). The stressful experience of bereavement increases noradrenaline in many people, and research shows that cancer metastasis rates are higher in bereaved people compared to similar nonbereaved people. There are many other factors at play, of course, and I am triangulating data that has not all been collected in the same study (which would be more compelling evidence), but the sympathetic nervous system may be a link between bereavement stress and cancer recurrence.

The researchers who discovered a clear pathway linking some types of stress and cancer growth are dedicated and careful scientists working in the field of psychoneuroimmunology. From the combined work of these pioneers, a new theory has emerged explaining how stress might affect our physical function, and it can be applied to bereavement stress as well. The psychoneuroimmunology theory I describe in this chapter is called the *conserved transcriptional response to adversity* (CTRA), developed by Steven Cole at UCLA in the 2010s. This theory posits that a stressful event changes our immune system. This occurs because our immune system responds to a stressor in our environment by trying to predict the best course for protecting us. This CTRA theory adds new information to our understanding of the interwoven mind-and-body response to bereavement, rather than supplanting earlier theories we have been discussing. CTRA adds to both attachment theory, which focused on regulating our physiology after the *loss* of the part of "us" embodied in our loved one, and cognitive stress theory, which focused on bereavement as an *added* demand for our system to cope with, resulting in our demands outweighing our resources. The CTRA theory suggests that our immune system may be altered by the loss of a loved one and function differently as we cope with these stressful aspects of grief.

Our immune system is always trying to protect us. One way it does this is to try to predict what immune challenges we might face at any given time. Our immune system then modifies itself based on that forecast, prioritizing those aspects of our immunity that will help us cope with the predicted challenges. For example, sometimes it is more likely we will have to cope with bacterial infections, and at other times we might be more likely to encounter viruses. It may surprise you to learn that our immune system uses different immune proteins and receptors to cope with bacteria as opposed to viruses. Thus, our body uses evidence of increased adversity (like bereavement stress), signaled by increases in our body's noradrenaline production (and our fight-or-flight behavior), to make a shift in the balance of our immune system's priorities. In the simplest terms, our immune system is constantly balancing production of different components for our body's needs.

This shift in the balance between antibacterial and antiviral responses includes changing what proteins our genes produce and which types of immune cells are preferentially produced. Think for a moment like you are the immune system. Thousands of years ago, stressful events likely involved being attacked by an animal or another person and were often followed by tissue damage to our body. Our immune system appears to have evolved to predict that when our stress is high, it should pivot toward protecting us from bacteria and other microbes that enter through wounded tissue. This means shifting the production of immune cells and proteins toward greater antibacterial inflammation, as I described after a cut on your skin in chapter 2.

This must have been effective as a prediction, because evolution seems to have conserved this shift toward antibacterial vigor when we are under stressful circumstances. But when facing stressors in our twenty-first century environment, like unemployment or discrimination or anxiety, a pro-inflammatory immune response is not helpful. Chronic inflammatory shifts in gene expression are not protecting us from being wounded or bacterial infection, but inflammation does spur chronic disease, like cardiovascular disease or diabetes. This shift of our immune function also leaves us more vulnerable to viruses, because resources shift away from the antiviral functions of our immune system.[2] Let me explain how my own research, in collaboration with Cole and others, demonstrated that bereaved individuals with more severe grief showed this CTRA shift.

Conserved Transcriptional Response to Adversity

My colleague Harald Gündel and I jointly mentor medical students interested in grief research. These students at Gündel's university in Germany add to their medical training by coming to the United States to conduct research with me. Sebastian Karl, who pioneered the aspirin study described in chapter 1, was one of these co-mentored students. But our first co-mentored student was Christian Schultze-Florey. He wanted to investigate whether bereavement stress led to a shift toward chronic inflammatory gene expression, or genomics. Since I had already learned psychophysiology, neuroimaging, and immunology, the last thing I wanted to do was learn another system like genomics. But my desire to be a good mentor inevitably wins, and so I found great colleagues at UCLA who could help us embark on the genomics study Schultze-Florey proposed.

If one wants to look at the genomic materials of cells, the easiest cells to obtain come from a simple blood draw. Our blood is full of white blood cells (or immune cells), and like all cells of our body, they contain both our genetic blueprint (DNA) and the RNA that is being transcribed from the subset of currently "active" genes. DNA may sound familiar, and you can probably picture it as the double-stranded helix, a ladder snaking around and around. But RNA is a single strand. RNA is in this untwisted strand form so that a specific segment of our DNA can get copied (or transcribed) for subsequent translation into a protein by our cell. By reading the recipe encoded in the RNA strand, our cell knows which protein to make.[3] Not every recipe that can be made is being made all the time, as our body needs different proteins at different moments. Plenty of recipes in cookbooks on my bookshelf never get used, even though I own them. By drawing blood and testing for the RNA currently found in that blood sample, researchers can see a real-time snapshot of what genes are being transcribed and translated into proteins. This process of going from DNA to RNA to a protein is called gene expression.

In 2007, Steven Cole began publishing studies that looked at the RNA found in blood samples from a variety of groups who were undergoing stress, especially long-term stress, like poverty, caregiving, and chronic loneliness.[4] He noticed a recurrent pattern in the RNA data: These groups of stressed people were expressing more inflammation-related genes and fewer antiviral-related genes. This was a great discovery: recognition that which genes were being expressed could vary as a function of stressful environmental conditions.

Genetic expression explained how stress could get under our skin, and our life events could change our human body and its functioning. Although science had documented more illnesses in people who were undergoing stress (like after the death of a loved one), there were few theories to explain what physiological mechanisms could account for this relationship between stress and illness. An increase in inflammation-related gene expression had previously been seen in animals undergoing social stressors, like mice who were on the losing end of fights for dominance. But Cole considered whether there was a bigger picture, using bioinformatic analysis to look at the big data produced by genomics. He grouped the RNA results he was finding into those genes that produced more in chronically stressed people according to their *transcription factors*.[5]

Inside a cell, transcription factors determine which genes get transcribed based on signals from the body outside the cell, signals such as hormones and neurotransmitters, like noradrenaline. As I said before, which genes are getting transcribed into proteins changes all the time, based on what is needed by our immune system. Through these transcription factors, the social environment outside of us, and our noradrenaline or cortisol response to it, can influence our body's gene expression. A stressful environment can shift the proteins being made toward greater inflammation.

Cole found that the group of transcription factors that told the genes to express more in chronically stressed people—called upregulation—pointed toward sympathetic nervous system and inflammatory factors. On the other hand, the group of transcription factors linked to lower gene expression in stressed people—downregulation—included antiviral (interferon) factors. This observation led Cole to consider whether the immune system was preparing itself. It seemed to be ramping up the proteins that would be needed for an antibacterial response to protect the body during this period of high stress, when stress meant our skin tissue being wounded in a fight, even at the expense of an antiviral response. Observing this large-scale pattern of changes in those undergoing chronic stress, you may now be able to see why Cole named the theory the (evolutionarily) conserved transcriptional response to adversity (CTRA).

It's Not the Loss, It's the Grief Response

Christian Schultze-Florey and I approached Steven Cole to see if he thought the CTRA could explain some of the health effects of bereavement as well. With the support of two other colleagues at UCLA, Michael Irwin and Otoniel Martínez-Maza, we began taking blood samples from the participants in my widowhood studies. By isolating the genetic material from the immune cells in the blood, Schultze-Florey was able to identify which RNA was being transcribed. Importantly, he was able to compare those who were widowed with similar older adults who were not bereaved. My own insight was the knowledge that most bereaved people are quite resilient. As difficult as the death of a loved one is, most of us experience the protest and despair of grief and yet find a way to restore a meaningful life. But a small group of bereaved people experience prolonged grief, the persistent and disruptive grieving trajectory even years after a loss. For this reason, I wanted to separate the widows and widowers into those who were experiencing prolonged grief and those who were experiencing typical grief, and compare each group separately to the nonbereaved group. From my prior work, I knew that those with prolonged grief have higher cortisol than more resilient bereaved folks, and this was another reason to compare the prolonged grief group with the nonbereaved group separately from comparing those with typical grief and the nonbereaved.

This grouping into three categories proved to be a clarifying lens through which Schultze-Florey, Cole, and I viewed the data. The widows and widowers with prolonged grief showed the CTRA pattern we had hypothesized—they showed downregulation of antiviral-related genes. Notably, the resilient typical grief group actually had a stronger antiviral genomic response.[6] These data convinced me that those who responded to loss with prolonged grief disorder really were having a different experience from those who did not develop prolonged grief—not just psychologically but deep in their cells as well.

This was a small initial study and needs to be replicated in larger and more diverse samples, but I feel confident in the results because they match the pattern of immune changes found across many types of chronic stress, not just bereavement. Importantly, this reduction in the antiviral arm of the immune system might account for the increased rates of flu and pneumonia mortality we see in bereaved spouses.[7] It might even account for the failure to become fully immunized against the flu even after receiving the flu vaccine, which we see in the data for older widows and widowers.[8] These data also offer an opportunity—the possibility that interventions targeting the stress response of bereaved people might affect not only their mental health but their physical state as well. To be clear, however, studies examining whether stress management interventions change the link between bereavement and medical outcomes still need to be conducted. Later in this chapter, I will describe a relaxation intervention study from my own lab that proved helpful for widow(er)s, but a definitive study would need to show all the links in the chain. In the same bereaved person, do we see decreased grief and stress levels, lower noradrenaline, higher expression of antiviral-related genes, and less flu and pneumonia? The grant funding needed for such a long and detailed study would be huge, and so for now, scientists infer this chain of events from the separate studies that show data in separate cordoned off sections of the proposed chain of events.

The Stress of Being with Those Who Are Dying

When I was in college at Northwestern University, I volunteered at Evanston Hospital as a "candy striper." Evanston Hospital coined that phrase, named for the red-and-white striped uniforms their volunteers wore during World War II. The candy striper program teaches young people how to come into a hospital environment, how to use medical precautions and regulations, but also how to walk into a patient's room and start up a conversation. When I moved to graduate school in Tucson, I did volunteer hospice training at Carondelet St. Mary's hospice, learning how to make a bed without moving the patient out of it and how to use foam "popsicles" of water to moisten patients' lips. More importantly, one of the older hospice nurses taught me to simply sit and listen to patients and family members. Perhaps in some way I was trying to figure out how I could have better supported my mother during her long terminal illness.

As many hospice volunteers will recognize, although I could provide calm and generous support to dying patients and their families, I had not responded effectively to my mom's fear of dying, her depression and anxiety, her expectations of me to soothe and comfort her. When I was hired in 2012 at the University of Arizona and moved there from UCLA, Al Kaszniak (my former dissertation advisor) suggested that I attend Being with Dying. This is a retreat program for hospice care professionals offered by Upaya Zen Center, a Zen Buddhist practice and training center in Santa Fe, New Mexico. Hospice work was meaningful to me but also emotionally exhausting. In my conversations with Al, I recognized he managed his emotions in response to overwhelming situations differently than I did. But I had no understanding of how he was able to do so. My hope was that there were skills I could learn from Buddhism, in a community of other palliative care clinicians, to keep from being overwhelmed by the dying process.

With all I have told you about my personality as a young person, you will not be surprised that I was terrified of one specific aspect of the Being with Dying program: The first two days of the five-day retreat were in complete silence. We listened to talks from program faculty and met with a small group of other attendees to discuss briefly, but otherwise, we did not speak to one another. Not during meals, not in the evening as we brushed our teeth in the dorm, not when we bumped into each other in the Zendo meditation hall as we bumbled through the Buddhist practices we were attempting to understand. Silence was again a revelation—both incredibly difficult for me and a window into a whole different way of being in the world. The profound impact of the program would resonate in me for years to come. Although I felt incredibly anxious and on edge throughout the retreat (and also excited by what I was learning and the others I met), I now recognize that this was only a window to possibility. It was developing the *habit* of sitting in meditation that would change my life, but this development would not happen for a few more years.

The insights most resonant for me from Being with Dying were the teachings of Roshi Joan Halifax about compassion, and how to develop the resilience to sustain dedication to working in difficult circumstances. Roshi Joan is a pioneer in the field of end-of-life care, having worked with dying cancer patients throughout her career, receiving numerous awards and honors from universities and healthcare institutions around the world. Her insight, distilled through years of teaching, distinguished two different motivations for caregiving, both by professionals and family caregivers.

We usually think of caregivers as motivated by empathic concern, an other-focused emotion that occurs when witnessing another's suffering. Empathic concern typically involves feelings such as empathy, compassion, and tenderness. But Roshi Joan pointed out there is another caregiver motivation discussed less often: the motivation to relieve one's own personal distress. This motivation also leads the caregiver to attend to another's suffering, but it is prompted by the desire to relieve one's own uncomfortable feelings when observing the other's distress. We want them to feel better so we can feel better. If the caregiver's arousal in response to witnessing the suffering of the other person is not regulated, that arousal leads to personal distress. Without learning to regulate the personal distress, the caregiver's arousal remains high. Over time, caregiver burnout can develop. But the caregiver can learn emotion regulation skills to reduce this arousal. We can learn to focus on our own body's arousal and remember to do so as we go through our workday. A focus on true compassion can also be a part of this emotion regulation, and meditation practices can help us develop skills for coming back to compassion as a viewpoint, again and again.

Roshi Joan described something I recognized all too well: an event that brings the caregiver into contact with another's distress. I thought of the many times I had walked into my mother's room at her nursing home as this type of event. All sorts of emotions were triggered: tenderness, sadness, guilt, remorse, frustration, empathy. Outside of my awareness, I would become more and more agitated. This overarousal is the fork in the road that we can become aware of with practice. One fork takes us down the path of self-focused reactions. Think of the fight, flee, or freeze reactions that we are innately wired for, which during caregiving can look like avoidance, abandonment, numbing, or even moral outrage, depending on the situation. Sitting on the edge of my mom's bed, I felt all of these reactions. Understandably, I wanted to get out of the situation, out of the room, and I was unable to tolerate myself inside my own skin. But these responses to my own internal state only extended my arousal, or at a minimum did nothing to help my mother, despite my huge-hearted feelings of compassion for her, scared and thin in her nightgown.

The other fork takes us down the road of emotion regulation. Roshi Joan created the GRACE model, an acronym to remind caregivers how they can respond differently when they recognize a triggering event or internal overarousal. This model is described beautifully in her book *Standing at the Edge: Finding Freedom Where Fear and Courage Meet.* Once you have **G**athered your awareness, you can **R**ecall your intention of compassion, the source of your motivation for what you will do next. By **A**ttuning to your own internal experience, and the experience of the person you feel compassion for, and the difference between these two states, you can **C**onsider what will serve best in the situation. Figuring out what will serve best often includes taking a different view of the situation, a perspective larger than your own experience or the experience of the other person—a broader third-person view looking down on the situation. Pausing, reflecting in a mere moment of silence, can lead to a decision about what is best for both people and for the relationship.

So often with my mother I believed that either I could get what I needed in the situation or my mother could get what she needed, but I never considered what our "us" needed, what our relationship needed, what could be done that would benefit us both overall. My exploration of both attachment theory and Buddhist interdependence has made me realize that she and I made up one system; we were both one and two at the same time. How she felt affected me, and how I felt affected her. My ability to calm myself influenced her anxiety and agitation, on the occasions I was able to do so. And this is the *E* of GRACE: Engaging in a compassionate and ethical way, attuned to the whole situation, and also being able to End the interaction when enough has been done, not persisting in ruminating about the stressful situation. There is always another present moment to become aware of in this rise and fall of events that humans marvelously experience as the flow of life.

What I eventually recognized was that it is natural and normal that we experience a rise in arousal, in noradrenaline, when we encounter distress in our loved one. It is natural and normal to want to help them, and we often have the skills to do so. But that rise in our own noradrenaline affects us, and can affect our health over a long period of caregiving. This rise in noradrenaline can happen during interactions in their hospital room, but for those of us who continue to perseverate on these thoughts and memories, the increase in noradrenaline can continue to happen even after their death. We may not be able to reduce our arousal while we are caregiving for a terminally ill loved one, but if we continue to have the same bodily stress response each time their illness or death comes to mind for years afterward, it makes sense that this will affect our health. We can learn to become aware of our distressing and grief-laden thoughts, and we can learn to relax our body and find calm in our mind—this is why grieving can be an opportunity for healing.

Viruses Know We Are Stressed

The way our immune system fights viruses changes when we experience stressful loss events. The intelligence of viruses makes it hard not to be fascinated by them. Viruses can persist in our body for our whole life, like the way the chicken pox virus can erupt as shingles decades later. These viruses remain dormant in our cells for years, but what makes them reemerge? Why do they wreak havoc at one specific moment, after the opportunity has been available for so long? One implication of the CTRA theory is that viruses can detect and emerge in the moment when our stressed body is most vulnerable. From the perspective of the virus, this is when the bodily environment is most favorable—when its chief adversary, the antiviral immune response, is reduced in response to noradrenaline and the shift to an antibacterial immune priority.

From the perspective of a virus, determining when one's host is stressed might be as simple as monitoring for high levels of noradrenaline. Researchers think many viruses have evolved to "listen in" on the translation of our psychological stress into our biochemistry, to detect the specific transcription factors that convey noradrenaline signals to the cell where viruses lie dormant. We have some evidence to suggest that the stress of bereavement is just such an opportunity viruses might be looking for to mount a comeback. Careful research by psychologists Angie LeRoy and Chris Fagundes at Rice University investigated viruses that are usually dormant (or latent) in people who had experienced the death of a spouse. Epstein-Barr is one of the many types of herpes virus, a type that most of us carry. We think of herpes virus as associated with cold sores, but Epstein-Barr virus does not have an outward manifestation we can see. In order to detect it, a blood test can demonstrate specific antibodies that the immune system makes to keep the virus under control. This was exactly the test that LeRoy and Fagundes used to determine if Epstein-Barr virus was replicating in widows and widowers, requiring more antibodies and thus resulting in detectable levels in blood samples.

In 101 participants widowed in the three months prior to this study, 99 of them were carrying the latent Epstein-Barr virus. This is similar to the portion of the general population who carry the virus. The bereaved participants were also asked about twelve different physical symptoms they might be experiencing, like headache or numbness and tingling. LeRoy and Fagundes found that the level of antibodies to the Epstein-Barr virus accounted for 21 percent of the variability in the symptoms people reported, even when they accounted for factors like insomnia, smoking, and other medical illnesses.[9] Bereaved people often describe a wide array of physical symptoms, seemingly with no particular cause, and this study suggests that these physical symptoms are not imagined but may be related to viral reactivation and the immune response. I will note that most bereaved people see a resolution of these symptoms over time. Most of us have a robust and effective immune system, after all, and so our body responds to the reactivation of a virus by creating more antibodies, resulting in quiescence of the virus once more.

Of course, in my dream research project, higher levels of sympathetic activity (higher levels of noradrenaline in the body) and reactivation of the Epstein-Barr virus would be assessed at the same time and in the same bereaved people. We would expect the virus activity to be higher not only in those with the most severe grief symptoms but also in those who are showing the most activity in their sympathetic nervous system. Theory suggests that noradrenaline is the message used by the virus to determine that this is the best time to break out and start replicating. The virus does not know how "stressed" we are, in terms of the subjective answer we report, but it may pick up on co-occurring noradrenaline in our cells where it hides.

The Tumor Microenvironment

Beyond the role of noradrenaline in waking up latent viruses in the body, there is another pathway through which bereavement stress might affect long-term medical outcomes, like cancer progression and metastasis. When we think about cancer, we usually think of an abstract diagnosis, but a tumor resides in a tiny world of cells. Solid tumors grow in organ tissue, like breast or liver. In fact, cancer cells start as normal cells in our tissue that begin to grow out of control, an error in the usual restraints on growth that keep our cells under control when they divide. Researchers have become interested in what the microenvironment of cells around the tumor might tell us about how the tumor lives and grows, once it has become a separate entity from the organ.

For example, if a tumor is going to grow faster than the organ tissue around it, the tumor needs a bigger blood supply to bring oxygen and nutrients to it. Communication between peripheral nerves and tumor cells also increases. Studies by researchers Steven Cole, Susan Lutgendorf, and Anil Sood have shown that chronic sympathetic nervous system activation can lead to increased noradrenaline levels in the tumor microenvironment.[10] They had the novel idea that if a mouse with a tumor was put under chronic stress, the pathways from stress to tumor growth and metastasis could be mapped. A mouse can be stressed by containing it in a small space. Although the mice are safe, multiple days of this kind of close confinement for a couple of hours a day constitutes chronic stress. This work led to the discovery that increased noradrenaline in the tumor from this chronic stress experience led to tumor growth, primarily through beta-adrenergic receptors on tumor cells. In an important extension of this finding, beta-blocker medications also reduced the effects of chronic stress on tumor growth.

Like adrenaline, noradrenaline can be released by the adrenal glands into the bloodstream, affecting the whole body. Or it can be released from tiny sympathetic nerve endings threaded through the tumor. Cole, Lutgendorf, and Sood discovered that although this confinement stress increased noradrenaline both in the bloodstream and in the tumor, blocking the noradrenaline in the bloodstream did not prevent the tumor growth. In other words, the sympathetic nerve endings in the tumor are where the signals increasing tumor growth are coming from. Understanding the exact pathway by which stress affects cancer progression enables us to target this mechanism. I have high hopes that someday bereavement stress can be better understood and treated, altering its potential to have medical consequences.

Bereavement stress is not different in many ways from other chronic stressors that cause changes in the body—there is no grief-specific protein that our body makes. The point here is not that grief has a unique physiology. The point is that grief includes a physiological response. People undergoing the chronic stress of bereavement are having physiological reactions, along with emotional and cognitive reactions. Grieving people may show different physiological functioning than those who are nonbereaved. Additionally, those who are having the most severe grief reactions seem to be dealing with more dysregulated immune physiology than more resilient bereaved people. Knowledge of the specific pathways from this stressful bereavement event to immune changes to medical problems can give us windows into places to intervene.

Intervention could happen through changing our thoughts about the stressful event with relaxation or meditation practices, or perhaps even through pharmacologically reducing noradrenaline production or blocking noradrenaline receptors. Intervention might include assessing and treating those with the highest levels of grief severity who may be at highest risk. For example, we might treat pneumonia in the bereaved earlier and more aggressively, like Lizzie from chapter 3 needed. It is worth making the point that all of these appointments and interventions take time, effort, and energy. Bereavement-leave policies are largely designed to offer time to make funeral arrangements, but the workplace could also educate and encourage healthcare targeted toward bereaved employees, perhaps preventing medical events and improving long-term health.

And Now for Some Good News

Habitual relaxation practices can influence the way the brain regulates our sympathetic nervous system, as my mother experienced.

In turn, research has shown that stress-management interventions may buffer us from poorer immunity, even at the cellular level, even in the face of significant stressors like cancer treatment.[11] A study from my own lab investigated mind-body interventions in bereavement. Psychologist Lindsey Knowles, a graduate student of mine at the time, proposed a study of mindfulness training for older widows and widowers. Mindfulness training develops a systematic practice of focusing one's attention on moment-to-moment experiences, emotions, and thoughts—importantly, while accepting whatever arises in the mind and not judging it.

Knowles wanted an active comparison intervention and chose progressive muscle relaxation. Progressive muscle relaxation training was methodized in the 1970s and is a practice of developing awareness of bodily tension and psychological stress, skill in muscle relaxation techniques, and attention to pleasant sensations of relaxation. The practice involves tensing a muscle group (like your forearms) and then relaxing them, followed by tensing a different muscle group, one group at a time in a specific order until you have worked through your whole body.

Although progressive muscle relaxation and mindfulness meditation are two different interventions focusing on different aspects of the present moment, they are similar in important ways as well, and both types of training can be done in a group setting over six weeks. Notably, Knowles wanted to include a third arm of the study for comparison as well. She asked some bereaved participants to join what we call a wait-list control group. This group does not participate in the active interventions, but they do all the same evaluations throughout the same time period. This allows us to compare how grief changed for people over this time without any intervention, since we know typically bereaved people show some improvement in grief symptoms naturally. For an intervention to be considered effective, it needs to show more benefit than the simple passage of time.

As you can see in the following figure, participants in both the mindfulness training and progressive muscle relaxation groups showed significant reductions over time in their grief symptoms.[12] But in the analysis comparing the groups, to our surprise, the progressive muscle relaxation group showed a faster rate of decline in grief severity than the wait-list group, and ended up with much lower grief symptoms at the one-month follow-up after the intervention ended. We had expected the mindfulness training group to improve more, but that was not the case.

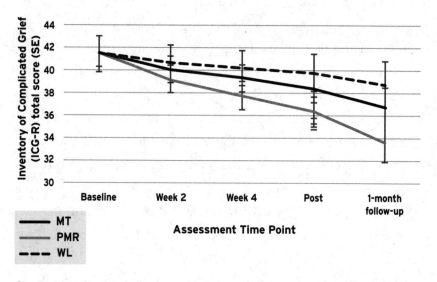

Change in grief severity over time in the two intervention groups (mindfulness training (MT) and progressive muscle relaxation (PMR)) as compared to the waitlist (WL) group that did not participate in an intervention but did the same assessments at the same time points as participants in the intervention groups.

Why would we see improvement with these two interventions? Both mindfulness training and progressive muscle relaxation focus on present-moment awareness, and practicing this shift in one's attention may reduce habitual ways of thinking, especially if this shift helps people relax. We could think of progressive muscle relaxation as a *body up* way to achieve this shift, and mindfulness training as a *mind down* way to achieve a similar state. In addition, progressive muscle relaxation may reduce the burden of physical distress associated with grief, addressing an aspect rarely considered in grieving.

What I take away from this study is also practical, because bereaved people can learn and practice progressive muscle relaxation through freely available online and app-based guided exercises. Community health workers and anyone in the helping professions can lead progressive muscle relaxation training groups without specialized training,[13] or with self-guided training.[14] Progressive muscle relaxation could be a beneficial add-on to existing bereavement support groups. This makes progressive muscle relaxation easier to implement than mindfulness training interventions, because best practices suggest that mindfulness-based interventions are better led by those with specialized training.[15]

Cancer Is Not Your Fault

Simply telling ourselves we *should* feel less stress will not affect our immune system. What matters is whether our sympathetic nervous system activity is reduced. This physiological change might co-occur with feeling calmed and soothed, but the point is that we cannot necessarily detect a change in our sympathetic activity. We certainly cannot detect how we are feeling if we do not take the time to sit and listen to our internal state. In order to know if our body is calm, we have to get to know what our body feels like in all sorts of different states of arousal, to get to know whether we feel relaxed or not. When I first started spending time sitting quietly and listening to the silence, I had no idea what my body was doing, having not spent much time focusing my attention on my internal state. But if we practice this sitting and listening over time so that we have a background picture to compare to our present moment, we can become more aware of the rise and fall of our internal arousal. Most of our physical state is below the level of conscious awareness, like our cortisol level, but we can raise our awareness of some aspects, like our heart rate, as we will discuss in chapter 9. In fact, even when we learn to affect our physical state through intentional relaxation, achieving this state is not possible all day, every day. My point here is that much as we may like to believe that we are, we are not in control of our body.

I am going to repeat that. We are not in control of our body, although we may be able to influence it. This is similar to the idea that we cannot force ourselves to fall asleep, but we can create the conditions that make falling asleep more likely. I mention this because my mother believed a myth that arises out of the data showing relaxation interventions can affect our physiological stress and physical health. A logical fallacy results from reversing this statement. The fallacy is that we can "cause" cancer or other illness by being too stressed or not thinking positively enough. This is simply not true. The fact that the logic works in one direction does not mean reverse causation makes sense. You cannot cause cancer with your thoughts or with shallow breathing, and if you have cancer, that does not mean it is your fault or that you failed to do something that would have prevented it.

Although stress management interventions give us an opportunity to potentially affect an ongoing process, the body's experience of stress comes from mechanisms well below the level of conscious awareness and outside of our control. Affecting our physical state through intentional practice can only change a small proportion of our experience each day. In fact, all illnesses have multiple causative factors, including genetics, the environment, and our history during development. Stress management interventions are still a worthwhile effort, as they may make life more bearable and they can have downstream impacts on our health, but the opposite idea, that we caused our own illness, is not accurate and produces greater distress for those who believe it.

Unhooking

In contrast to my experience of my mother's terminal illness, my father was not afraid of dying, and this was his gift to my sister and me. Only a couple of months after I attended Being with Dying at the Upaya Zen Center, I got a phone call that my father had suffered a heart attack. This did not come as a surprise; in my father's later years, his body stopped making red blood cells, necessitating regular blood transfusions. The downside of these life-giving transfusions was a buildup of iron in his body, which we knew would eventually weaken his heart.

By the time I walked into his hospital room in my hometown, I had already entered many hospital rooms to care for him. In his final decade of life, I had flown to Las Vegas when he became dehydrated and delirious during a Kiwanis convention. I had flown to Montana after a slow bleed in his aging brain necessitated neurosurgery. But when I walked into his hospital room this time, in the final week I would spend with him, I had a very different perspective, a different set of skills that I had learned from Buddhist practices. Spending time with my mother during her terminal illness was overwhelming, and although I knew that my anxiety and internal agitation were not helping, I did not know what else I could do in those moments with her. I did not know how I could change my response to the heart-wrenching situation of my mother's distress. But I had learned so much before I faced my father's final days.

Mindfulness meditation is a practice that allows the mind to rest in the present moment. In that moment, there is only that moment, although in future moments you may need to plan or consider or do a million things to care for someone who is dying. But in this moment, there is only the sight and sound of what is before you. There is continually remembering to return to the feeling of your breath, the rise and fall of it, when you discover your mind has flitted away to all the details you need to take care of. Mindfulness practice allows for the calm at the eye of the storm, without pretending the storm does not exist. I recognized that not only did I need this practice to face my father's death but he needed it as well.

I arrived first at the hospital after my father's heart attack, knowing my sister would be there only hours later and feeling the great relief that we would be together in our care for him. I was able to sit with him, to hold his hand—so familiar with its loose skin, brown age spots, and round nails—so similar to the shape of my own hand, but in a larger, man-size version. I was able to meet his eyes, offering him simply my presence. I had no magic drugs, but I could feel his worry and physical tension drain away. I was there. I would be there no matter what. I loved him. He knew it and felt supported in making his unknown journey ahead. By feeling my own calm, I could stay there, with nothing else I needed to do or even say, not compelled to flee or freeze or fight my internal state.

Of course, in the days that followed, I often lost this sense of calm—I needed breaks from the sheer weight of what was happening. I appreciated the ready glass of wine from our dear family friends with whom my sister and I stayed, the opportunity to spill my worries and stories about the events of the day. I needed to hear their memories of caring for him, reassuring my sister and me that our father mattered and belonged in our community. Dying often takes longer than we expect, even when someone has stopped eating and drinking. The natural process of shutting down a human body has many steps. I have come to think of this process like Velcro—disentangling each of the myriad individual hooks that attach us to life, unhooking each of our physical systems, unhooking from each of the relationships that make "us" who we are. This takes time, work that seems to be done in the semicomatose state many slip into, although we know so little about their internal experience. I was able to be present and witness his passing, defending his need for peace and quiet when others seemed not to understand, stroking his forehead for comfort, talking with my sister because he probably could still hear that we were in the room. My father's was a peaceful death, not only peaceful for him but deeply peaceful in me as well.

Perhaps feeling the meaningfulness of caregiving, despite the difficulty, and leaning on those to whom I felt close also had an impact on my immune cells.[16] Research has demonstrated that many factors affect our stress levels for good as well as bad and, in turn, affect our health at a cellular level. For example, social support or a greater sense of purpose in life appears to affect gene expression, reducing inflammation and enhancing antiviral defenses.[17] Perhaps this helps to account for why those who feel they are restoring a meaningful life during bereavement do not show some of the same health issues that plague those who continue to struggle with protest and despair. Importantly, however we handle the loss of our loved ones, grieving is an opportunity from which we can learn more about how our physical body is a part of our closest relationships, encoding our love as well as our stress, adding to the accumulating wisdom of a life fully lived.

PART TWO

Healing During Grieving

CHAPTER 7

Energy and Motivation

Grief is the natural reaction to loss; in the case of the death of a loved one, it is also a loss of a part of yourself, the "us" part of yourself. It is a loss of a part of yourself that helps you to function in the world. Of course, you must learn to cope with the waves of grief that can unexpectedly overwhelm you. But as the dual process model by Maggie Stroebe and Henk Schut (introduced in chapter 3) points out, coping with loss also means dealing with the stresses of restoring a meaningful life. *What will retirement look like for me without my deceased spouse, if we had always planned it together? How can I survive the pressures of college if my lifelong cheerleader, my mom, has died?* Bereavement can significantly change our goals, and our energy and motivation for those goals, so that we barely recognize ourselves. In this chapter, I will share why psychologists think these responses happen during grieving, why energy and motivation are different from each other, and how I learned to keep putting one foot after another with day-to-day challenges to both.

Protest and Despair

John Bowlby, in his observations of attachment behavior of infants separated from their caregivers, and Myron Hofer, in his observations of rat pups separated from their mothers, distinguished two different responses to loss. They called these two responses *protest* and *despair*, which I introduced in chapter 2. But energy and motivation function quite differently in these two different responses. In infants, protest behavior is designed to motivate reunion with one's caregiver. Protest involves searching and calling out, in order to find the caregiver or attract their attention. We recognize these responses in adults who have loss as well. For example, the visceral experience, the panic, that most of us are having as we imagine looking down to find our toddler is not by our side—this is protest. The feeling is full of energy, motivating us to find them. That motivation comes from our attachment neurobiology in the brain, which responds when we become aware we are separated from a loved one.

I have described protest as "*Oh no, they're gone?!?*" Protest includes awareness that the loved one is missing, which makes it distinct from utter disbelief. Protest means you are fully aware they are gone, like in the grocery store example. On the other hand, I describe despair as "*Oh no . . . they're gone . . .*" Despair embodies withdrawal, giving up, with full awareness of the gravity of loss. If you imagine feeling despair right now, notice how the energy of protest has drained away. If the evolutionary advantage of protest was to give you the energy to quest for your mate, parent, or child, then despair transitions you to give up on the search. There is an evolutionary advantage for this shift to giving up, which may point to why despair is also a natural response to loss. Running around, searching for your one-and-only makes you more likely to attract predators, and if your one-and-only is going to be gone for a long time, conserving energy by withdrawing becomes more important.

But despair also serves a cognitive function on the learning curve of accepting the painful reality of loss. Despair *enables* you to give up the costly search of protest, expensive for the body in all the cortisol and adrenaline it requires. Despair has a function, and its function is to give up a fruitless search. I know that very few of us think of despair as being worthwhile or useful. For this reason, supporting in grieving people both the experience of protest and the experience of despair is important.

Of course, despair is not the end of the story, because despair also carries hopelessness. We cannot yet imagine that we could ever live a meaningful life, given the loss of this important person. It is okay to allow a grieving person to express protest, to express despair. This is very difficult for us to do—as individuals and as a society. We value the perfect meal, the happy wedding, the social media posts of cute pets and gorgeous sunsets. But if despair has a function, if it enables us to accept the reality of the loss (which must ultimately happen in order that we adjust to our new world), then accepting that grieving people will feel despair from time to time may be necessary.

So in addition to supporting a grieving person in their current emotional experience, support means lending them your hope, while not negating their current pain. Caring for bereaved people includes both supporting where they are now and encouraging them for the future. Hope can be shared despite the fact that neither of you can yet imagine what the future will look like. But the hope that a full life will one day be possible again reflects the empirical fact that the vast majority of grieving people do find a way to adapt. Although despair contains the understanding that the loved one is truly gone, it also contains the erroneous belief that one's life will be filled with despair forever. How could it not, since the loved one is not coming back? But adaptation means adjusting to a changed reality, which includes grief but will also include joy and love again. Restoring a meaningful life constructs it *around* the fact of loss, with the full range of human experience, with pain and compassion and joy and love.

Here I will mention again that protest and despair are not stages. This is not a stage model. Protest and despair responses do not happen once and then they are done. In fact, for people who cannot tolerate the feeling of despair, they may return to a protest response as soon as the irrevocable nature of the despair overcomes them. Others cannot tolerate the inkling of reality that accompanies protest, and retreat to disbelief or avoidance, other responses we may have when we are aware of our loss. Disbelief is the feeling that the loss is just not real. This is distinct from avoidance, or the active attempt to keep the reality of the loss out of awareness. Because grieving can be thought of as a form of learning, the realizations of loss must painfully dawn on us again and again. Sometimes we can experience all of these responses in a single conversation with a grieving person, bouncing back and forth between protest and despair, perhaps even with some disbelief and hope sprinkled in. Grieving is exhausting, both for the griever and the grief adjacent!

It's Not Just Bereavement

The death of a loved one is the most obvious cause of grief, but grief does not only occur when a loved one dies. Because our loved ones are as vital to our survival as food and water, our brain evolved specific mechanisms to keep us with them, our attachment neurobiology. The death of a loved one is the most universal of losses that all social mammals face. But the breaking of an attachment bond happens in many other situations for human beings, including breakup, divorce, estrangement, or the empty nest. These experiences of grief are obviously not the same, but our brain may use similar neural machinery to understand this change in the original bond it encoded.

We understand less about the neurobiology of grieving during other losses that do not involve a relationship, like the loss of a job or the loss of health. But the grief over the death of a loved one and the grief over other losses are connected by the fact that almost all grief involves the loss of a part of our self. For example, I think of myself as a daughter, but the word *daughter* implies two people in the world. The word *spouse* implies two people, or even *best friend*. And yet, I use the word *daughter* to describe myself, an attribute of who I am, and it informs how I function in the world. But what does it mean to be a daughter if I have no parents? Grieving means figuring out what it looks like to be a good daughter, without parents in the world. Others must learn what it means to be a widow, different from being married or single, or what it means to be a parent after a child has died.

When we look at other types of losses, losses that are not due to the death of a loved one, we can see the similarity with the loss of a part of ourselves. When I retire, I will have tremendous grief over the loss of my job, because being a professor is a huge part of how I function in the world. When I was diagnosed with MS, I faced the loss of health, the loss of functioning in the world as my usually energetic self. These losses created enormous grief for me. My grieving, my learning curve, following my MS diagnosis mirrored closely what I have observed in others' grief over the death of a loved one. My grieving included coming to understand my physiological reaction to loss and how this knowledge became an opportunity for physical healing. I hope that describing my experience from the inside will resonate with the many experiences of grief you may have felt.

With any permanent loss, protest and despair are likely human responses. With loss of health, I protested and continued to attempt things I was not able to do any longer, draining me further. With retirement, we may despair as to whether we are still a valuable member of society; we may be unable to imagine how being an elder could be valuable in and of itself. Once we digest the idea that protest and despair are typical responses to loss, we can recognize them in many situations. I remember a good friend who was about to be married and have a child. She found herself completely annoyed with needing to get a new car: She knew that her zippy little MINI Cooper was not going to fit a car seat, but she could not understand her reaction—why was she so upset? I wondered aloud with her if she was grieving the loss of her single self, able to take off at a moment's notice, no need to account for whether anyone else would be inconvenienced. She was relieved, as this explanation rang true for her. She was protesting the loss of the freedoms of singlehood through her annoyance over having to sell her beloved car, even while simultaneously wanting to marry and start a family. Our conversation gave her permission to be sad, to reflect on the real loss that she was incurring, of all she had known and enjoyed about this part of her life, the loss of more than just a car. Having accepted that this was grief, and no longer perplexed by her protest feelings, she was able to turn her attention to the joys of her new life, while not ignoring her need to hold on to the memories and aspects of herself that were so familiar and enjoyable.

Shifting Out of Protest and Despair

I have a vivid memory that I look back on often, a turning point for me in understanding my own reaction to the loss of my health. It had been at least five years since my MS diagnosis. I was standing in the hallway of my home, and I was overcome with panic. I was overcome with the image that my body was like an astronaut's suit. Without oxygen pumped to me and wastes removed, without the suit's arms and legs allowing movement, I realized I could not survive. I would die if something happened to my suit. My breathing became fast and shallow, a classic panic attack. I could not shake the disembodied feeling that my body was separate from me, and the equally distressing feeling that I was wholly dependent on it for life. But for the first time I was able to also experience a moment of despair, without returning to the panic. Perhaps the years of practicing meditation kicked in. Perhaps my years of studying grief and death anxiety filtered in. For whatever reason, I went from feeling overwhelming terror to feeling the calm of despair that accompanies the draining away of panicking protest. I was going to die someday.

Although I had awoken every morning for several years with this hopeless thought of my own mortality, my usual reaction to the horrifying thought had been to throw myself out of bed and into work for nine hours. This time, the protest drained away, and for that moment, I just felt the gravity of the reality that I was going to die someday. Not today, but someday. It was so sad—I loved life, I loved my work, I loved my wife and my best friend and my sister. I did not want to lose all of that. I could not comprehend the mystery of what it would mean not to live. But the truth of my own mortality was firmly planted in my mind. There was nothing I could do to prevent my eventual death; someday my body would not work as my astronaut suit on this Earth. This time, the panic in response to that knowledge evaporated.

This deep sadness and despair felt different from my depressed moods. This despair felt clean and uncomplicated, and I felt it more deeply in my bones than the diffuse feelings of depression. Along with the sadness and uncertainty about death, I had a fleeting moment of accepting the awe-filled thought that my death was part of the circle of life. That my life was also part of all life, and all life arises and passes away, the way each breath we take fills our lungs and leaves us. I had an inkling of accepting my mortality. Of course, this calm insight did not mean I had no more moments of panic and despair and depression. I continued to respond to my existential fears by overworking, but acceptance of the clarity and simplicity of the time-limited nature of my existence remained in the back of my mind, deep in my cells somewhere, from that day forward.

Years later, it dawned on me how much our fear of death contains our fear of separation from loved ones. Death anxiety is—at least in part—separation anxiety, perhaps at a very basic neurobiological level. Following that day, whenever panic and death anxiety arose (like sitting on a plane during turbulence), I would think, *You're not wrong. You are going to die someday. You only have now. Are you okay with that? Is there anything you should be doing differently so that when you do face that moment, you have done and said and experienced all that you want to?* Over time, my answer became *Yes, I am okay with that.* It is the reality, that one day I will face my last moment. There is no point in pretending I will not. And my life is full, my loved ones know they are loved, and I have left on this Earth books and lectures about what I learned during my time here. So while I do not want to die, I am okay with the idea that I will. Moreover, needing to be okay with my mortality motivates me every day to live in the present, and resolves so much of the terror and panic and anxiety I carried around for so long. Despair resolves protest, and accepting resolves despair.

Motivation and Energy

This vivid insight about my panicky stress response to fear of my own mortality was a turning point for me, and as I have said, grieving can be a turning point in life that allows an opportunity for physical healing. But insight is not enough. Years of practice, and other insights, were needed to shift my understanding of how to live a meaningful life with my loss of energy, to understand how to use my motivation to achieve my life goals while accepting my loss of health. I simply could not continue to ricochet through my life, protesting the reality that I could not do everything I wanted to. I had to take up the practice of regulating my breathing and heart rate, reducing the physiological toll of anxiety. I had to give up busyness as a way of coping. Reactivity and ricochet contrasted with the calm *responsiveness* I eventually learned and (mostly) am able to engage. Responsiveness integrates a deep awareness of how I am feeling in the moment with the bigger picture of what I can do that will feel meaningful. It focuses on what I can do to create a meaningful life, given I cannot know how long that life will be. Learning responsiveness and regulation meant I had to cope with occasional overwhelming waves of loneliness, of separation anxiety. Grieving people may face, for the first time in their lives, incredibly strong emotions like outrage or crushing guilt. Learning to recognize, understand, and manage waves of strong feelings is a key to changing the way we use our finite energy. And one of the first things I had to learn was how to disentangle my energy from my motivation.

Motivation is our interest in engaging; energy is our capacity to engage. Defining them so neatly suggests these aspects would be easy to assess for oneself, but my experience is that they are deeply intertwined. For example, when I have low motivation, I feel I cannot possibly go to a social event or to the gym. And yet, if I overcome motivation's dissuading message and engage anyway, my energy often buoys me along after all, developing into a productive writing session or a good lecture. High motivation, on the other hand, may obscure my low energy. For me, this is a dangerous combination, because not having the energy to back up my motivation means I am depleted quickly, and makes me question whether I should have tried the activity in the first place. This depletion makes me second-guess myself the next time I might try to do something.

The depression duo—lack of motivation and lack of energy—create the hardest days of all. Even if I manage to overcome my lack of motivation, I discover I cannot do anything productive after all. In these moments, I have to dig deep and remind myself that this too will pass, and I am not flawed or lazy or unlovable. This has taken me a long time to learn, and sometimes I still fail to attribute the lack of capacity to my current temporary state, rather than to a deep, permanent inner flaw. A bereaved person may similarly have to overcome the belief that they will never again be able to live a meaningful life, or that they are too weak to cope with their grief. Rather, we can learn the skill of self-compassion, which requires practice, returning our focus to a kind stance toward ourselves. We can see this variation in our energy and motivation as a part of our authentic selves, and being authentic brings a kind of peace all its own.

All this introspection has also led me to savor, to celebrate, the days I have both energy and motivation. It is such joy! On these days, I continuously return to awareness of my body and mind working as the miraculous instruments they are. I make note of these days, a reminder to be saved up for the days when this is not my experience. It does not mean I have to do everything on these days that I cannot do on other days—I have come to realize that it is more important to do a reasonable amount and enjoy this capacity. I also make note of these good days to my current partner, a reminder that the days when I complain that I feel like I am walking through Jell-O are temporary, the mysterious inner workings of this person he loves.

Pacing

Some people experience a restless, agitated energy during bereavement, and others experience exhaustion, a deep tiredness that makes them not want to answer the phone or get dressed. Regardless of the physical causes of fatigue, we still must cope with it. How are we supposed to restore a meaningful life for ourselves if pouring cereal counts as making dinner? Part of accepting our grief includes accepting (at least for now, for an uncertain period of time) that we may not be able to function as our previous selves, and yet life is still happening. My experience of coping with fatigue from chronic illness may be useful to you, and that is the topic we take up now.

I do not want to understate the impact of deep fatigue. Pulitzer Prize–winning journalist Ed Yong quotes a woman with long COVID in his article "Fatigue Can Shatter a Person." She describes fatigue as "a complete depletion of the essence of who you are, of your life force."[1] We forget the complete intertwining of our body and our sense of who we are. We *are* what we *do*, and where one ends and the other begins is fuzzy. Our sense of self is built on the familiar experience of our daily interaction with the world. Exhaustion can mean we do not know who we are anymore; we can feel *This just isn't me.*

I had the enormous fortune when I was first diagnosed with MS to have access to the Marilyn Hilton MS Achievement Center at UCLA. Part of the MS Center is a workout room, where physical therapists help you learn to exercise with your newly fragile body, reducing the fear that you will harm yourself. I had never been a person who exercised regularly. At the MS Center, I was taught that exercise was important for maintaining my physical reserves but also an important aspect of mental health. In fact, meta-analyses have shown that exercise can be effective for treating depression.[2] But how do you work out when you are exhausted? Our newly diagnosed group would work out for ten minutes, overcoming our low motivation. Then we would assess whether we felt more fatigued than we did before we started. If our fatigue had increased, we were encouraged to stop. But if we felt the same, or even energized, we were encouraged to continue. Over time, we gained experience in listening to what we could do on any given day and the effect that working out had on us later that day and later in the week. I find that exercise helps my mood and energy in the longer term, but it can deplete me in the short term, even for mental work. Even today, I tend to work out later in the day, when I have few commitments beyond dinner and TV viewing.

Paced exercise requires lots of trial and error. With a personal trainer, I discovered that thirty minutes was feasible when I was religious about including one-minute timed rest periods throughout, although half an hour had been an endurance level I had thought was impossible. I must constantly evaluate my energy in real time, while remaining acutely aware of the consequences of what I choose to do. This remains a challenge, but a challenge that has led me to understand my body more deeply than I had ever thought possible.

I am amazed when I can do something that seemed impossible a few weeks earlier, and I am patient with myself when I must give up after ten minutes. At some very deep level, I feel like I know who I am (for now!) and this strengthens my sense of self in every situation I confront.

Pacing may be a key strategy for you as you face varying levels of energy from day to day during grieving. If your deep exhaustion leads you to less active days, which may affect your ability to sleep and eat as well, then consider whether bereavement is an opportunity to get to know your own body's signals better. Accepting your body's state, at least today, while also asking if you can work out for ten minutes or going for a stroll at lunch, demonstrates care for your grieving body. If you pay attention (the topic of chapter 9), feedback from your body during these ten minutes gives you more knowledge of who you are, what feels good and what feels bad, and how to move through the world as the bereaved person you are now.

Plan B

The days, weeks, and even months that follow the death of a loved one may feel like a blur. In many cases, one's schedule changes dramatically. For example, if you were caring for a loved one prior to their death, a huge hole looms in the day, previously filled with phone calls, running to the pharmacy, inquiring about new symptoms, managing appointments, and on and on. Or, if the death was unexpected, filling in for all the things that your deceased spouse did becomes a daily obstacle course with unpleasant discoveries and frustration, in addition to caring for your children and managing your job. For any death, the amount of administration, dealing with insurance companies and lawyers and banks and the coroner's office, is overwhelming. All of this must be faced without the one person you would usually lean on in stressful situations. No wonder most of us feel so bad.

Even when the hubbub starts to recede, or the reality sets in, there is dealing with the people who know you have had a loss and do not know how to relate to you, and the people who do not know and you must tell. Coupled with fatigue, your desire to withdraw may become your modal response. The uncertainty of how you will feel on any given day makes it difficult to make any plans. How can you accept an invitation if you have no idea whether you will be able to manage putting on clothes and shoes next weekend?

For this challenge of planning activities, another thing I learned at the MS Center might be helpful to you: Think of your energy as a bank account. It is a limited resource, and if you spend it on one thing, there will be less there for something else. Budgeting decisions must be made based on what is worth the investment. Cereal might be a fine choice for dinner, if you have the energy left to pay bills or get your hair cut. But as I looked at the pie chart depicting my energy expenditures during the early months of my MS diagnosis, I realized that I had stopped spending any energy on socializing. I had just cut that expense out almost entirely, prioritizing work and chores and family responsibilities. Faced with low energy, I felt that I could not afford to spend it on a phone call with an out-of-town friend, going to dinner with a much-loved neighbor, or a birthday party for a close colleague. Once you picture a pie chart of your energy expenditures, you can imagine that you might take a little slice from each category, spending a little less energy on work or chores in order to leave inner reserves for social activities that may also renew your soul.

In addition, you might consider a plan B—the option you create in case the arrangement you have made turns out to be too difficult when the time comes. For example, if you would like to go to a concert with friends, you might ask if they would also consider a plan B: If your energy is low that day, would they come over and just watch a movie instead? If you are invited to a holiday party, you might tell them that you probably will not be able to stay, but you would love to drop off flowers or a gift and say hello. You can always stay if you end up feeling more festive when you arrive than you had imagined. There are several advantages to using a plan B. By making a plan B in advance, you feel less guilty when you are not able to engage as you would like because you feel tearful or drained. It also sets expectations; others understand that either option is a possibility. Figuring out a plan B in advance is helpful because you can think it through more easily than when you already feel drained.

If you are close to the people with whom you are making plans, a plan B can become a way to talk about how things are going for you, what is working and not working in your life right now. One of the challenges of restoring a life for yourself, a life with the absence of your beloved, is that you want to develop a life that is enjoyable, meaningful, and full of loving relationships. But you may not yet have the energy or social motivation for that life. If you continue to decline invitations, however, you may find that you get invited less, and that is probably not the life you would like to live eventually. Being honest about your situation, or even just letting people know your plan B without discussing why, means that others can feel you are still available even if you are the mourning version of yourself. This is often more encouraging to others than you just avoiding them or continuing to cancel plans. And you may surprise yourself on some of those occasions.

No One Said This Would Be Easy

You may learn to respond to waves of grief or low energy by focusing on managing who you are at the moment, but needing to manage so carefully in and of itself is frustrating and sad and hard. I sometimes think, *Yes, I can manage it, but I wish I didn't have to!* Frustratingly, people will say, "But you look so good . . . ," which is irrelevant and can feel invalidating, although it is not intended that way. People probably tell you that they feel tired too, or they have grief. The fact that they feel tired does not change the fact that you feel tired. Facing grief is not a competitive event. There is only coming to know your own experience, deeply and fully, and learning how best to move forward today, given that your experience is true. Sometimes I think, *Wow, I feel so irritable today*—and I allow that to influence my decisions about whether to attend a meeting or simply not respond to an email in this state. Neither of these actions changes the reality of how I feel, but they are choices we can make in response, given the truth of the moment.

Making these choices is a practice, the practice of learning to be a person who has grief, an opportunity to heal one's body by responding to its needs. In Yong's article in *The Atlantic*, he quotes a woman who says, "It takes so much self-control and strength to do less, to be less, to shrink your life down to one or two small things from which you try to extract joy in order to survive." For her and perhaps for you, rest has become both "a medical necessity and a radical act of defiance."[3] The healing is centered in one's own experience; it is the facility to be and do with less energy, with more awareness and compassion. In truth, we will all face the loss of energy at some point, if only when our body ages beyond our expectations. I sometimes feel I just got a jump start on understanding the aging process, and feel fortunate I can use all I have learned to navigate all sorts of hurdles that are the nature of being human.

From our Puritan and pioneer and immigrant heritage, Americans believe at a deep, unconscious level that productivity equates to worth. Even mental health is sometimes couched as whether someone can "function," which is often mistaken to mean whether someone can work, rather than more accurately being able to do the things that a person finds meaningful. Other challenges can add to the grieving depletion. Energy-limiting conditions, including auto-immune disorders and depression, affect more women than men, and women have to deal with the second shift and often do more to prove themselves in the workplace.[4] Gwynn Dujardin, an English professor at Queen's University in Canada, points out that women in literature such as Circe and Dido were judged harshly for offering the temptation of rest to Odysseus and Aeneas, taking them off course in their epic quests.[5] So it is not an exaggeration to say that learning to rest, and to value rest, takes determination. It took me many years to admit, first to myself and then to others around me, that I needed a nap every day. It makes me more productive, but I feared the impression that I was lazy or weak. The withdrawal of grieving is usually for a shorter period of time than a chronic illness, and of course, we should always consult a doctor to make sure there is not an underlying cause of fatigue that could be addressed. But if rest is needed, then working it into your every day may have longer-term benefits to your health, both mental and physical.

It's Okay That You're Not Okay

We forget that our modern cultural focus on the need to prove to ourselves and others that we are okay looks so different from how grief has historically been expressed. As author Megan Devine says, it's okay that you're not okay.[6] It is okay to take some time away, to withdraw for a time, to direct our attention to all we are feeling and thinking.

In September 2022, I happened to be in London when Queen Elizabeth died. The depth and breadth of the grief response to the queen's death was astounding, even with my awareness of her beloved status. I realized that this was not just the individual responses of many people. It was a throwback to an era of how grief used to be expressed, albeit with greater collective attention because of her status.

I was fascinated by how that attention was maintained—all the advertisements at bus stops and on the Tube were switched overnight to memorial photos of the queen. Large department store window displays were replaced with memorials, while smaller shops dressed their window mannequins in black. The pub where we had Sunday lunch showed the memorial coverage on every TV, as opposed to typical wall-to-wall screens of football and rugby. I noticed people on the street with their cell phone cameras angled high into the sky, and as I followed their gaze, I saw the thirty-seven-floor BT Tower with a photo of the queen circling the full circumference of its video display. This change made me aware of how much of our attention is constantly being directed as we walk through a city, and how a national response could still (at least in the UK) redirect that attention to something of societal importance like death. I did not think a contemporary culture was capable of focusing on one topic for seven days, but for a whole week, BBC One ran programming only about the queen's life. When American friends asked why British television channels stopped running ads the first night after her death (showing only the channel logo for the entire three minutes each time instead), I tried to explain that commerce seemed disrespectful or even frivolous at a moment when a death was the focus of our attention. Grieving needs time and space and support, and this was historically a collective response, not just an individual one.

As I watched this collective outpouring of grief in Britain, directing the attention of a whole country to the role of ritual to cope with and acknowledge mourning, I recognized that it enabled us to transcend our daily lives with recognition of the pause grief deserves, and with an opportunity to come together. For days, people stood in a five-mile-long, fourteen-hour line, awaiting their turn to pay their respects to the queen. The stories they told to one another in the line and to reporters about what she meant to them demonstrated grief as a collective sharing, beyond just a collective emotion. Perhaps much can be learned about the importance of this coming together for mourning, when the death of a loved one happens in our own family or neighborhood. This is universal, human, collective grief—taking time away from daily life, creating physical expressions of mourning in laying flowers, and most of all, our joint attention on that which we have lost. What we learn from reflecting on grief shapes both the personal narratives of our lives and the narrative of our community.

We Can Do Hard Things

Protest and despair are characteristic responses to loss, but they are not the ultimate integration of the loss into our ongoing life. Neither protest nor despair result in the peace that can accompany accepting the reality, giving up fighting the truth of the loss, moving beyond the idea that the loss means we can no longer do or be anything of value. Loss is part of life—this life we were born into with these human bodies and human relationships. Protest and despair are natural responses to loss, reactions of our brain and body that we cannot control. I had to realize that my fear of dying was causing me enormous physical tension and anxiety. Accepting the loss of my health, and even my eventual mortality, gave me back energy that I was spending without realizing it.

We can do hard things, as we will see in the next chapter. For a time during grieving, we may need to do the hard thing of stepping back. Perhaps we are in a community or friend group that allows us to do this, or we can find the person who *gets it* and can support us in this challenging shift in priorities. We may not be able to control the natural grief reaction, but over time and with practice, we can learn to cope with the fact that we have grief. Building a life around our loss means learning how to navigate our need to withdraw or our low energy level. We can learn to shift our perspective, and we can develop new and skillful habits, which enable us to experience all the aspects of our wondrous human life, the pain and the joy. Best of all, with coming to know uncertainty and loss, we find we are more deeply connected to others who also face loss, who also *get it*. We may see how we all are but a single instance of shared human loss and grief, giving us a reason to feel more connection between and compassion toward others and ourselves.

CHAPTER 8

Healthy Habits

When I got divorced, I no longer had to compromise about what I cooked or ate, when I went to bed, or what my early morning routine looked like. But I quickly discovered it is not easier to cook for one person, and without shared bedtime routines, my own schedule seemed to dissolve. We think of ourselves as independent creatures as adults, but in truth, our day-to-day activities are bound up in the interdependent activities of our relationships.

This interdependence can be seen in evidence of the co-regulation patterns of hormones that regulate our sleep, energy, and appetite. A study by psychologists Dora Hopf and Beate Ditzen at Heidelberg University documented this during the COVID pandemic: People who had a partner showed better cortisol regulation than those who were single, even if they did not live with their partner.[1] Without the input of our loved one, we must find a new equilibrium. After a loss, changing our daily habits means we must acknowledge that our loved one is gone and live in a way that reflects our new reality. This may seem too hard to do, or even impossible. But without changing our habits, we are constantly struggling to interact in the world in an automatic way, a way that no longer fits how the world *is*. Changing our habits takes time, but knowing that grief and the difficulty of acknowledging the depth of our loss is why we feel so off-balance may lead to better planning, and to self-compassion as we find a new path.

Just as a cast supports a broken bone by providing structure around the healing break, we can create routines and schedules that help us to develop new habits. These serve as resources even when our motivation or energy is low. For example, I might ask myself, *Why am I taking a shower?* Not because I have the motivation (the lack of which I am ignoring), but because it is morning and that is next on my schedule. As you might imagine, the most basic habits that regularize our physical functions, like eating and sleeping, are the most important.

The Nights Are Hard

Diana Chirinos is a psychologist at Northwestern University Feinberg School of Medicine who studies insomnia in widowed people. Her findings show that those who have the greatest rumination—the thoughts that go round and round in one's head—also have the greatest difficulty sleeping.[2] When we have had significant loss, we often dread going to bed, because with nothing to distract us, our thoughts turn to grief. For widowed people, the bed is usually a strong reminder of loss. One of Chirinos's participants commented, "I don't like sleeping in my bed . . . It's funny how you form habits—I'm still on the edge of the bed, the whole bed is empty. I could spread out, but I don't."

Chirinos is also studying how to treat insomnia in bereaved people, and part of her motivation is that research shows sleep medications are not effective. For example, in a randomized study in London, bereaved people were prescribed twenty pills of either low-dose diazepam or a placebo, provided within two weeks of the death of their partner and to be taken as needed across the following six weeks. This mirrors the way sleep medications are often given to bereaved people by general practitioners. From baseline to the end of the six weeks, more people in the diazepam group had trouble getting to sleep in thirty minutes or less after going to bed, but fewer people in the placebo group had trouble getting to sleep.[3] This seems counterintuitive. After all, why would sleep medication not help improve their sleep? The answer may be related to what is called the 3P model of insomnia.

This 3P model is a general explanation of how sleep problems can arise, and not just during bereavement. It focuses on three conditions that must come together for us to develop insomnia: predisposing, precipitating, and perpetuating factors.[4] The first *p* is for *predisposing* factors that increase the risk of developing insomnia, and in the case of bereavement, these are the factors present before the death of a loved one. They include genetically determined factors, personality traits, and environmental situations that are already in place when a loss happens. For example, a long period of caregiving prior to a loss is a predisposing factor that can increase insomnia during bereavement. The second p is for a *precipitating* event, in this case the death event. Loss often triggers initial sleep difficulty that initiates short-term insomnia. It is well documented that most of us experience initial sleep difficulties in bereavement. In a systematic review, eighty-five different published studies described these grief-related sleep problems.[5] Most important is the third p, because these are the *perpetuating* factors. These are the things that maintain insomnia once we have begun having difficulty sleeping because of our loss.

Perpetuating factors are also things that we may have some control over, unlike the things that happened before the loss or the fact that our loved one has died. Essentially, some of the things we do to cope with common, expected sleep problems during bereavement may perpetuate the insomnia, turning it into a long-term problem. For most people who have sleep difficulties during bereavement, normal sleep patterns will return. Only a couple of studies examine the natural course of sleep difficulty in bereavement, but one Dutch study followed recent widows over time. The widows showed significantly more sleep difficulties at four months than a nonbereaved group of women. Overall, the widows returned to their baseline sleep patterns after a year, although their sleep was also significantly better by seven and ten months after their loss.[6]

So, what coping strategies for insomnia backfire, perpetuating our short-term sleep difficulties and turning them into long-term insomnia? Now we can unravel the mystery of why prescribed diazepam did not help insomnia in the study of bereaved people I described earlier.

The physiological co-regulation we share with our bonded loved one means that when our loved one dies, we must learn to regulate our system without this relationship, previously a part of our own functioning. Learning to live with their absence includes adjusting our sleep system to go to sleep, to stay asleep, and to wake up without them there, or even simply without the comforting knowledge they are somewhere in the world. Grieving is partly learning to regulate our physiology without the input of their warmth, their smell, their touch, or their texting us good night. Human beings are resilient, and our sleep system will adjust, even though it takes time and is frustrating and we are often tired a lot. But if we introduce medication as a part of that adjustment, our system is adjusting to the loss by adopting those chemicals as a part of our sleep system. This means that we are regulating with the input of those medications. We will still need to adjust to functioning without them, eventually. If we are taking the pills intermittently, as in the London study, our sleep system has difficulty knowing what to expect and how to adjust on the nights we do not take them. This is why sleep medication can ironically be a factor perpetuating insomnia.

Introducing medication or other substances to help us get to sleep, or to keep us alert during the day, are not the only perpetuating factors that can turn insomnia into a long-term problem. Particular beliefs about sleep and specific behaviors have been proven to maintain insomnia. These include using the bed for anything other than sleep or sex, like watching TV or using a smartphone. We want to condition ourselves to the idea that getting into bed means sleep will follow. Similarly, spending a long time in bed unsuccessfully trying to get to sleep or to return to sleep weakens the bed-means-sleep conditioning. If you have insomnia, and you are taking more than fifteen minutes to get to sleep (or back to sleep), the recommendation is to get up and do something quiet in another place until you feel sleepy again (and preferably without a lit screen in front of you). Have you ever lain in bed, so annoyed by the fact that you cannot sleep that it is impossible to relax? Some people become more upset and anxious about their insomnia over time and develop unrealistic expectations about sleep, such as the idea that unbroken sleep throughout the night is required to feel rested, or they underestimate the amount of time they are asleep, or they believe they will not be able to work the next day if they are tired. Having these thoughts during the night paradoxically make it harder to fall asleep, as we get more frustrated because of the outcome we anticipate the following day.

Most important as we learn to reregulate our sleep system is to keep a consistent schedule. Getting up in the morning at the same time, even when we are tired or had a poor night, is key. Getting exercise or movement in your day (but not right before bed) and refraining from daytime naps (including falling asleep in front of the TV in the evening) are vital to our sleep system. Although I usually take a nap every day, if I am experiencing a bout of insomnia or when I have jet lag, I skip my nap to help consolidate my sleep to nighttime hours.

But the most important thing I can tell you is that cognitive behavioral therapy for insomnia (CBT-I) is a proven effective treatment. In head-to-head comparisons, CBT-I is effective and does not interfere with long-term sleep function like sleep medication can. CBT-I is available (even as teletherapy) from board-certified specialists.[7] You may read a list of sleep problems like the one in the previous paragraph and believe you already know these things, but it is very different to have a specialist identify your particular perpetuating factors and explain which techniques will ameliorate them. There are mathematical formulas for how long to spend in bed given your specific sleep patterns, and the best half hour of the day to nap if you feel it is necessary. These formulas, techniques, and tailored interventions can be gained only through seeing a professionally trained clinician. I cannot encourage you enough to get behavioral treatment for insomnia if you are suffering from it—it can be a turning point in healing after loss.

Benefits of Treating Insomnia Beyond Grief Therapy

There is an additional reason to seek treatment for insomnia, even if you already have a grief support group or are in therapy related to your grief. First, in a large study I did with Dutch colleagues Maud de Feijter and Annemarie Luik, and Mayo Clinic's Brian Arizmendi, we discovered that having insomnia prior to the death of a loved one increased the chances of developing prolonged grief disorder.[8] So not only can grief cause insomnia but already having insomnia can also make grieving harder. Second, several studies have now documented that even after people with prolonged grief disorder have successfully completed psychotherapeutic treatment, poor sleep may persist.[9] That is, even though feelings of yearning have decreased, and new skills for coping with waves of grief have developed, insomnia can remain an issue. The good news is that following CBT-I treatment, 100 percent of bereaved people said they would recommend this brief intervention for guidance in how to restructure their sleep–wake schedule.[10]

Sleep abnormalities have health risks of their own, even unrelated to medical consequences of bereavement. Surprisingly, although sleeping too little on average (less than five hours) is associated with poor health, sleeping too much (eight and a half hours or more) is also associated with poor health. One way researchers assess good health is to consider the combined health of many physiological systems together, including cardiovascular, immune, lipid-metabolic, glucose-metabolic, sympathetic, parasympathetic, and hypothalamic-pituitary-adrenal systems. This is sometimes called the multisystem biological risk index. Colleagues from UCLA, including Michael Irwin, Jude Carroll, and Teresa Seeman, have shown both sleeping too few hours on average and sleeping too many hours on average are associated with poorer multisystem functioning of the body.[11] Data for this finding comes from a longitudinal study of over one thousand participants, called the Midlife Development in the United States (MIDUS) study.

While too many and too few hours of sleep reported in the study were both associated with poorer physiological regulation, the number of hours we sleep is distinct from sleep quality. Sleep quality is how well we feel we have slept, a subjective sense of how restorative one's sleep was. Compared to a normal sleep duration (between six and a half and eight and a half hours), all short sleepers reported poor sleep quality. But in long sleepers, it was only those who reported poor sleep quality who showed multisystem biological risk. For example, I would be considered a long sleeper—and I have been a long sleeper all my life. I do best when I get about nine hours of sleep. This is not necessarily a bad thing for my health, because subjectively I feel all my hours of sleep are of good quality. However, if you are in bed for eight and a half or more hours but are not getting good sleep during this time, this may be a health issue and worth your seeking treatment.

What to Eat

Eating is another behavior completely thrown off after our loved one dies. Many bereaved people lose weight, and widowed adults eat more meals alone, have more snacks, and make fewer home-made meals than their married counterparts.[12] After my mother died, my father was famous for inviting all the widowers in my hometown over on Sundays for dinner. He and I usually talked on the phone on Sunday after he went to Mass, and he would tell me what was on sale at the grocery store that week, leading to his delicious chicken or pork chops made in a slow cooker. He cooked basic and tasty meals for the widowers, leading many of them to learn about the effort and attention it takes to develop cooking skills, and that combining cooking and eating with socializing can make it more pleasurable.

Marriage is associated with better health for men, like my father's friends, but we know it has an impact on women as well. In a longitudinal study of over 80,000 women, those who became widowed or divorced lost weight, compared with women who remained married.[13] Women skipped regular meals after losing their spouse or partner, and ate fewer vegetables relative to women who stayed married. Similar patterns are seen with other health behaviors: married women also engaged in more mammogram screening, exercise, and smoking abstinence. However, the future of widowed women diverged, depending on what happened next in these women's lives. At a four-year follow-up, women who had remarried regained a healthy weight, as compared with women who had remained unmarried. Married life may bring regular meal patterns and an increase in food intake through social interactions, although it is not the only way to develop new eating patterns.

You may think that losing weight is a good thing for health, but bereavement as a cause of weight loss is not usually healthy. Eating patterns change in bereavement for people of all ages, but nutrition plays a major role in the health and longevity of older people. In fact, weight loss is often a sign of poor nutrition. Older people who lose weight are less likely to gain it back compared to younger people. In a study in Pittsburgh, fifty-eight older adults widowed in the past six months were matched with the same number of older married people. Weight loss was greater in the first year of the widowhood, even when adjusting for baseline weight.[14] Widowed participants lost about two pounds on average, but some widows lost up to ten pounds. Not everyone lost weight after the death of their spouse— higher cognitive functioning protected against weight loss, pointing to the idea that learning to shop, cook, prepare, and eat meals is a demanding task in the new reality after significant loss.

Learning What Works for Us

When we look at women who have been widowed for more than three years, compared to more recent widows, we see that most women find their way to establishing new patterns.[15] This fact points to the resilience of bereaved people but also points out that we should strengthen support for those who have difficulty during this transition. How does the transition in eating patterns usually work? In-depth interviews by Canadian researcher Elisabeth Vesnaver led to the discovery that it is a two-stage process, in which women first fall into new patterns and then establish a personal food system that works for them.[16]

Making decisions over a lifetime leads to the development of personal mealtime preferences. But in addition, couples have complex negotiation about which foods and how much food to eat, and about preparation and timing of mealtimes, made over the duration of their relationships. Caregiving for a terminally ill loved one can also shift these decisions and habits, and you may find that meals have become a significant source of stress. Whatever your history, bereavement can be a shift in the social context in which you eat.

In Vesnaver's study, she found that widowhood generally resulted in a "lifting" of food-related obligations as a wife, the most common (though not universal) role in couples. Widowed women reacted either by enjoying the opportunity to eat according to their personal preferences or with a disorientation due to the loss of these obligations (a feeling that reflects my own experience after divorce). Recent widows described less desire to cook for themselves and consequently skipped meals or made less healthy choices. But longer-term widows found that they had adjusted and their eating situation had stabilized.

The first stage after the death of a spouse or partner in Vesnaver's study included disruption of all the different activities associated with making food and eating, and included a lack of interest in food in general. Some study participants also fell into a pattern of snacking as a form of coping, which they recognized as distinct from consciously adopting a method of grazing as a way to have small meals throughout the day. Grazing, when used as a choice, tended to replace a three-meals-a-day structure, as opposed to interfering with it. Snacking, on the other hand, was recognized as a form of coping as related to their mood, often to fill long afternoons or evenings spent alone. In general, widows viewed this first stage as temporary, and it did not match their values around health and self-care. This disruption stage sometimes overlapped with the beginning of the second stage of establishing their new food system.

In the second stage, there was a growing recognition of the misalignment between their eating behavior and what they valued. They described a realization of how much their new eating patterns differed from their previous patterns, or even from social norms around meals, sometimes from comments by others. The system that each person ultimately developed was unique to themselves and their situation. Some women relied more on frozen prepared food, in response to a dislike of cooking elaborate meals. Others focused more on the social side of meals, growing networks that allowed them to eat with others or developing a relationship with a "family" of staff at a local restaurant.

Many participants described a watershed moment, when they decided they needed to take charge of their lives, including physical needs but also linked to emotional and even financial health. My nan described a moment just such as this after my grandfather died when she was still a young woman. She would tell the story of sitting on the side of the bed and thinking, *Well, you can sink or you can swim. And you are going to swim.* She became a *dinner lady*, as the British call it, at the cafeteria at a local school. In Vesnaver's study, the watershed realization led to an adjustment in how the widows approached meals over time, leading from a system designed to satisfy the needs of the couple to a system for obtaining and enjoying food themselves.

Loneliness

Bereaved people in Chirinos's focus groups identified the role of loneliness in their struggles to cope with sleeping, and Vesnaver's participants identified the role of loneliness in their struggles to cope with eating. Loneliness is a significant issue in grieving, although it is distinct from yearning for the loved one specifically. What I mean is that when widowed people have friends or adult children with whom they feel close, they feel less lonely. But the lack of loneliness does not reduce their yearning for their deceased partner. Loneliness and grief over a one-and-only are distinct, but loneliness is a problem that can emerge in the context of bereavement.

Loneliness is different from simply being alone. As one of Chirinos's focus group participants said, "I'm by myself in this apartment now, and so I'm just kind of dealing with . . . It's so quiet at the end of the day." That subjective feeling of loneliness is distinct from objective social isolation, as not everyone feels lonely even when they are alone, but isolation does increase the opportunity that loneliness will develop. Loneliness is very common in the first year of bereavement and is reported as one of the most distressing parts of grief. For most bereaved people, loneliness lessens over time. This points to the unique and challenging difficulty with chronic loneliness. By the same token, chronic loneliness tends to prolong grieving and contributes to greater depression.

Our American culture has developed beliefs about loneliness that may make resolving feelings of loneliness harder. Many negative feelings, such as hunger or pain, motivate our behavior to relieve that unpleasant state. These motivating feelings are often tied to survival needs, like food and water, but we sometimes fail to recognize that social connection is necessary for survival as well. From this perspective, loneliness is just a signal that we need to be with other people, the same way that hunger is a signal that we need food. We even know that the brain encodes these motivating feelings similarly, from a neuroimaging study that deprived participants of either food or social interaction for ten hours. When these participants were then shown pictures in a neuroimaging scan, the same midbrain regions were activated in hungry people seeing food pictures and lonely people seeing social interactions.[17]

However, in our culture, we often perceive our loneliness as a failure or a fault, instead of as information about our current state. The analogy to hunger here is helpful as well. For example, when I am feeling lonely, I have been known to sit on my couch and think, *I hate this feeling. Why am I so lonely? What is wrong with me? Why do my friends not call?* And yet, I rarely sit on my couch and think, *Why am I so hungry? What is wrong with me?* Instead, I simply get up and eat something!

Because social connection is an ongoing need for human beings, loneliness is predictable. Hunger is also predictable. Because we know we will feel hungry, we go grocery shopping in advance of feeling hungry. We even take the time to make a grocery list, to plan what we will need and want to eat in the week to come. But how many of us take the time to make a social connection list for the week? How are we going to get our social needs met, so that when we feel lonely we can remind ourselves we have something already planned to look forward to?

When a person notices they are lonely, and successfully seeks out others for connection, they feel better. But when a person does not successfully find social connection after feeling lonely, again and again, eventually they go from acute to chronic loneliness. We see also this shift in behavior in other social animals, not only in humans. Animals become more aggressive or withdrawn when they have been isolated, even when they are later offered the opportunity to socialize. The downside of chronic loneliness is that the motivation to reach out to others decreases when we go from acute to chronic loneliness. Chronic loneliness is harder to cope with.

When we are chronically lonely, research shows we pay more attention to negative aspects of our social interactions. We remember more of the negative aspects of previous social events. We have more negative expectations of upcoming social interactions. And because we have negative expectations, we are more likely to behave in ways that decrease social satisfaction when we do meet up with people—a self-fulfilling prophecy. Unfortunately, this also means that simply bringing chronically lonely people together may not result in enjoyment or new friendships. Research studies across a range of different loneliness interventions show that the interventions must address the negative thoughts about social interactions that develop in chronic loneliness. People who are chronically lonely think, *I'm not going to enjoy this. Those people aren't really interested in me. It's not worth my time and energy to make an effort to get to know these people better; they are likely to be a disappointment.* Or, *If they really wanted to spend time with me, they would call. I'm not reaching out to them if they don't call first. Why should it be up to me?* Interventions that specifically address these negative thoughts and predictions are empirically more effective than just increasing opportunities for socializing or even teaching social skills.

There are some things we can do to increase our ability to prevent acute loneliness from turning into chronic loneliness, even during bereavement, a time when we may naturally feel disconnected from others around us. Research by psychologist Kristen Neff demonstrates that showing kindness to ourselves allows us to feel worthy of being loved.[18] Volunteering engages us with others who have common interests and meaningful goals, which often make social interactions easier. Seeing others on a regular schedule (like a standing weekly coffee or lunch date) prevents us from having to overcome common social anxiety around making an invitation every time. Having meaningful conversations increases the enjoyment of social interactions, compared to small talk. (If after a conversation we know nothing more about each other than we knew before, then that was small talk.[19]) Finally, seeing people face-to-face, and without the distractions of phones or screens, is usually most rewarding. Face-to-face does not necessarily mean you have to sit across from each other and gaze into each other's eyes; it can mean going for a walk together, or any interaction during which the other person has your full attention. Social interaction creates opportunities for eating as well as stimulates wakefulness, which can help us to sleep at night. Spending meaningful time with others not only feeds our need for social connection but also helps to regulate our appetite and sleep.

Grief and Sex

One change that does not get considered much with grief is sex. And yet, I think this change in our experience can be hard for us to fathom. Some people experience a desire to have sex right after the death of a loved one (although I am unaware of any research on this topic), which seems so incongruous when one is feeling so much grief (at least in our culture). But we have to remember how much our hormones shift when we experience loss, and this might help to make more sense of that urge. People have sex for a variety of different reasons, and these may all affect our sexual behavior in the following months. Some people have sex because it is comforting, and I think this an obvious need for many in grief. Others have sex to connect with someone, a life-affirming experience in the midst of what can be a deep and existential distress. On the other hand, many people have no sex drive while grieving. In my own experience after the loss of my mother, I could not bear to touch into my inner experience of my physical body, which is usually a prerequisite to having sex.

On the other hand, I want to mention an experience recounted by the famous psychiatrist Irvin Yalom in the book *A Matter of Life and Death*.[20] Others might see this as odd, and Yalom found himself utterly bewildered by his own experience. In a chapter entitled "Sex and Grief," he describes obsessive sexual thoughts about women in general and women he knew after the death of his wife, with powerful and persistent images. He describes an overpowering desire and then intense shame, since he adored his wife and felt completely bereft. He had no desire to "move on" and he movingly describes his grief, his yearning for his wife. He even searches the scientific literature to find a reason for his sudden powerful sex drive, but finds no research on the topic. Although I also have no scientific studies to share, I think the hormonal shifts after loss, and the changes in the attachment neurobiology that occur, make these sexual thoughts and feelings seem more understandable.

A related experience in bereavement is touch deprivation, sometimes called skin hunger. Touch deprivation is exactly what it sounds like: the unfulfilled biological need for human touch and physical contact. I have heard this described by widows, but I think it is probably more widespread. Interest in this experience of touch deprivation increased during the COVID pandemic, because of the increased experience of social distancing and isolation across the world. A study in Italy investigated positive touch (i.e., hugs, kisses, caresses, holding hands) in almost a thousand people during the pandemic and found that the frequency of hugs with a cohabitating partner was associated with significantly lower symptoms of depression.[21] A similar finding was reported in the United States. The authors point to the pre-pandemic finding that frequent hugs between spouses/partners are associated with lower blood pressure and higher oxytocin levels.[22] Given that spousal bereavement shares the impact of deprivation from physical touch with a partner, the distressing experience of skin hunger may work through similar mechanisms.

The Interaction of Physical and Mental Health

When older bereaved people consistently monitor their caloric intake, this predicts their feeling greater energy. Sleeping approximately six and a half to eight and a half hours per night in the months after a loss predicts better social and emotional health.[23] I am struck that the questions a primary-care doctor might ask a bereaved patient could show a great deal of care and interest by genuinely inquiring about what changes have occurred in that patient's eating and sleeping. Doctors and nurse practitioners can provide good information about the normality of the disruption that happens after loss. They can offer reassurance that most bereaved people adjust over time, but also suggest a follow-up appointment to discuss these topics again if healthy patterns are not reestablished.

The interaction of physical and mental health is critical. Bereavement reveals that one's thoughts, feelings, behavior, and physical body are one system, and they interdependently affect one another.[24] By recognizing that the body and the mind do not exist in isolation, we can use our thoughts, our feelings, our behavior, and our physical body as levers to improve our ongoing adaptation and create healthier lives. We may wish with all of our might that this loss in our life had never happened, but the reality is that we are in a new life. We might as well put our mind to establishing new healthy habits in our changed life, despite it being an unwelcome opportunity.

Accepting How You Feel, Yet Doing What You Value

It is important to distinguish your situation from how you are *feeling* about your situation. In addition, it helps to recognize and legitimize what you *want* to do in response to your situation, but distinguish that from what you will actually *do*. For example, the situation may be that it is time to go to bed according to your planned schedule. You may be feeling apprehensive, not wanting to face the time alone with your thoughts without distractions, worried that you will feel grief and loneliness. All of these are legitimate feelings, simply the realities of your situation. Saying, "I shouldn't feel this way; that's silly," does not change that you feel this way. Awareness of how you are feeling is necessary to understand the new reality you are living. Saying, "I can't go to bed now because I feel this way," fails to make the distinction between what you want to do and what you choose to do.

Knowing that you do not want to go to bed is important information and can be used to strategize for the future. Perhaps you can engage in some problem-solving, because the reality is that you do not want to go to bed. A good friend whose husband died in their bedroom had family and friends come over and repaint the room, rearrange the furniture, and gift her new bed linens. She was able to associate going to bed with this comforting memory of support, rather than earlier distressing memories. Another friend began sleeping with a body pillow, which she found very comforting. Yet another friend began using a sound machine, because she could look forward to soothing nature sounds that reminded her of her childhood and helped her not to ruminate when she got into bed.

But problem-solving is for daytime, not for the moment when you are going to bed. You have established what the situation is and how you feel. You can still choose to follow a plan you created, even when you do not feel like it, but you need to have a reason. You might say to yourself, "I don't really want to go to bed now, but I know that in the long run establishing a good sleep habit is critical to eventually feeling better. I'm going to get up and go to bed, because *I want to feel better in the long run*, even if it means I need to overcome my inertia right now." Sometimes it is easier to call on your better self: "Future Mary-Frances wants me to go to bed, and I trust her. All right, I'm going . . ."

Especially for those who have shopped and cooked and planned meals for their family, they may have the skills to develop new eating habits but not be used to applying them to themselves. It may be useful to consider not *What do I want to eat?* but *What does Mary-Frances need to eat?* This slightly different perspective takes us outside of ourselves to make better choices from our wiser, ideal self. Even if one is choosing to eat snacks to accommodate a decreased appetite, purchasing from the perspective of one's wiser self may lead to making healthier choices.

This approach can be used with any new situation in which you find yourself:

Situation: *I have been invited to join a campaign for a cause I think is worthy.*

Feeling: *I am too tired to deal with people after work.*

What you *want* to do: *I want to ignore the invitation. I don't want to go. But I feel badly that I don't want to go after they've gone to so much effort, and it is for a worthy cause.*

Here is the key: *What I actually do is a reflection of me, of who I am, my values and goals and purpose. That is uniquely me, and no one else can know what is the right decision for me.*

You may decide: *I will go anyway, even though I don't feel like it, because my goal is to spend more time socializing.*

Or you may decide: *I'm just too tired, and I value my downtime, even among other values I hold. Even if they don't understand, I'm prioritizing my rest.*

Neither of these reasons are motivated by guilt or because of what others will think of me, but by my values, my purpose in approaching this situation.

Thoughts that arise around a choice you "should" make can lead you to the awareness that motivations for what you should do may be different from motivations for what you feel like doing. This self-knowledge helps to clarify who you are. Feeling this authenticity is a source of strength. You can make a choice once you understand your contrary feelings and motivations, and base those decisions *not* on what you should do, and *not* on how you feel at the moment, but on what you value most. This process requires time for self-reflection, and sometimes benefits from a friendly ear, from someone who is not giving you advice but is trying to understand your range of feelings. Avoiding situations, or avoiding your mix of feelings, does not help in the long run. Neither of these reasons are motivated by guilt or because of what others will think of *me*, but by *my* values, *my* purpose in approaching this situation.

Freedom from vs. Freedom to

I want to clarify that in acute grief, in the days and weeks after a loved one dies, our body is just reacting to the loss. Our initial reactions are not by choice, and we have no control over what our cortisol levels or motivation levels are like, over whether we cry or do not cry. In the long run, we learn to reregulate, but this is a slow and stuttering process. Regulation cannot be forced; we cannot will ourselves to sleep or to have an appetite. But we can set up the conditions that make new, healthy self-regulation most likely, just as we would wear a cast and use crutches to support the healing of a broken leg. Believing we are responsible for waves of grief that mean we do not want to go to bed or to cook dinner is like feeling responsible for our hair not growing faster or for our broken bone not knitting back together faster. Physical healing and growth, even from loss, are natural long-term processes. Yet we often do not even give ourselves as much time to grieve as we would give ourselves to heal from physical injuries.

In the months and years after a loved one dies, we may realize that we are struggling with the freedom of what to do next. Psychologist Erich Fromm described a distinction that resonated with me when I was grieving the death of my mother, in his aptly titled 1941 book *Escape from Freedom*[25]. When my mom died, I felt free from her needs, from my difficult interactions with her. But it was a long time before I realized that the challenge for me was how to manage the freedom of making my own choices about what to do, who to be, how to live. In many ways, I turned to the familiarity of responding to the needs of others, of making my decisions about what to do through what they needed from me. In the months after I divorced and my father died, I again found myself struggling to know how I wanted to live, with the burdensome freedom to do what I felt was the right choice for me. We are both free and responsible to create the life that works for us, within the constraints that we all face, including our appetite and insomnia and grief, and all the things that are true in our present reality.

CHAPTER 9

Awareness, Attention, Intention, and Purpose

I have mentioned that I am divorced, but what I have not said is that I had never lived alone before that time. Grief over the end of a relationship, whether from death or divorce, still requires figuring out how to operate in the world without our partner. We must discover what it means to be without the person we always thought would be there, even if the breakup was our decision. Grief is not only about the loss that happened; it is also about what life is like now, in the painful new reality. In my forties, after seventeen years of marriage, I found myself coming home to a condo that was silent. I hated it. I dreaded it. I feared it. The intensity of my anxiety when I was alone made me feel like I was losing my mind. How could an accomplished adult be so terrified?

One particular aspect of this experience made me concerned

about my mental state. When I came home from work, I had this bizarre experience that if I had no one to tell about the events of my day, it was almost as though they had not happened. I had the feeling that without talking about these little incidents with my ex-wife, it was as though they had not happened, almost as though I did not exist. This was a feeling I had never heard described in all my years as a psychologist. You can imagine my surprise when nine years later I read Irvin and Marilyn Yalom's *A Matter of Death and Life* and Irvin described a similar experience after the death of Marilyn, his wife. Irvin has decades of experience working with people as a psychiatrist. At the age of eighty-eight, he also found himself living alone for the first time. He wrote:

> If I see an outstanding TV show, I long to tell Marilyn about it, and over and again I have to remind myself that there *is* no Marilyn and that this TV show, this flake of life, is of value and interest even though Marilyn shall never share it. . . . I am *not* referring to loneliness. It's a matter of learning that something can have value and interest and importance *even if I am the only one who experiences it, even if I cannot share it with Marilyn.*[1]

My fear was not just about being alone, in the traditional sense of fear for what might happen to me. As with my younger self, I had trouble putting my finger on what was so scary, what made my heart race and dread rise. On many panicky evenings, I would go walking fast in my new neighborhood, trying to match my pace to my elevated heart rate. In a conversation with my dear friend and mentor Al Kaszniak, I finally blurted out, "It sounds crazy, but I'm afraid to be alone in my own mind." Fortunately, from his experience as a longtime Buddhist meditation teacher and his deep compassion, he told me that perhaps this was not so surprising, that many human beings had faced this sort of existential fear. With his reassurance, I realized that taking up meditation in a more disciplined way might allow me to face this fear, with the guidance of teachers and a tradition that for thousands of years had studied being alone in the mind. Of course, my strong impulse was simply to jump into a new relationship to quench my fear, but both my sister and my best friend explicitly forbade me to date again for at least six months!

Peace Can Be Found in the Present Moment

After my divorce, I had the good fortune to attend the Buddhist Society Summer School, a weeklong retreat held in the English countryside. Newcomers and experienced meditators come together, attend talks, and meditate several times a day. Being alone in your mind while many people around you are also being alone in their minds feels very supportive. I was reminded often of Quaker meetings and struck by the impact that many people—together, yet attending to their own minds—can have. The sense of listening was familiar, although now I had more of a sense that I was listening to my mind and body, the space in which my thoughts, my feelings, my sensations arose. Meditating together as a community is not the same as spending time alone, but developing an understanding of how my mind worked, and a decreased fear of it, enabled me to spend time with my thoughts and feelings in a new way when I was alone. Learning to focus on counting my breaths, and letting go of anything else that arose in my mind, seemed both extraordinarily simple and virtually impossible.

I hear many grieving people say, "I can't get rid of these thoughts." My own experience is that there is no getting rid of thoughts. Rather, we can discover that thoughts simply come to us, seemingly out of nowhere, and they also move on if we give them time and do not get too entangled in their content. We can see them as just thoughts, like individual clouds passing overhead or individual cars passing on a street. *Ah, there is a blue car. Now it has passed and there is another one. This one is red. Now I have a thought about my last argument with my ex-wife. Now that thought has passed, and I have a thought about whether my roof needs replacing.* By seeing thoughts just as events that arise in the mind, without giving them any special importance (*Oooh! I need to think about this! I need to solve this! I need to argue about this!*), it is easier to experience them being let go of. That seems like a strangely ungrammatical way to phrase that, but this is how it can feel. *I* am not letting them go, but rather, thoughts come and go in my mind all on their own. But I can help to create the conditions in which they do not stick around. Part of creating these conditions in my mind is to have a regular meditation practice, to over and over again experience thoughts being let go of when I realize I have become distracted by them. And again and again, return to counting my breaths whenever I become aware that I am not in the present moment.

I learned many things over the course of the summer school, but one in particular made a big impact on me. Buddhists emphasize the idea of daily practice. This means that attending to the present moment is not just for when you are meditating, sitting on a cushion. You can be in the present moment throughout your day, in any moment. Later I would also learn about the idea of *samu*, or work practice: meditation in the midst of activity. This is often carrying out tasks, like washing dishes or chopping vegetables, while staying completely attentive to the activity you are doing. Instead of focusing on your breathing, you simply focus on your intended activity, but again, while also letting go of anything else that arises in your mind. Of course, since this is virtually impossible at first, you simply practice returning to the present moment again and again, as you remember that this is your intention. It is easier to practice this with something simple like house chores, physical activities centered around concrete movements in the present moment, such as feeling the water move from the glass to the dish towel as you dry it.

Observing this continuous change is also why focusing on the breath is a good practice, because your breath is constantly changing in the present moment, a physical movement as your chest expands and then deflates. As I got better at staying in the present moment for chores, I realized I could try to apply this same *samu* frame of mind to my academic work—I could try to stay present with the laptop I was typing on, or stay fully awake to what I was saying in the present moment in a lecture, rather than lecturing on autopilot. This is much harder, as I use my mind for many things when I work (remembering, listening, generating ideas). But the continuous return to the present moment when I remember to do so makes whatever work I am doing more meaningful, more intentional. It also allows me moments of rest throughout my day, when I pause to refocus on my breathing, and dramatically changes the amount of energy I expend on fear and grief.

Your Energy Expenditure Is Probably Different

Grief's impact may skirt around our awareness, pushing at our heart and nudging our immune cells. Although grief runs amok in our subconscious, we may be able to raise the physical and emotional consequences of the loss of our loved one to consciousness. Grieving was an opportunity for me to learn that I was spending my energy on my fear of being alone, and avoiding any possible ruptures in relationships that might lead to my being alone. Other people as they grieve learn different things about how they are spending their energy. One young man discovered that his father had instilled in him a fear of being taken advantage of, that the worst thing in the world was being someone else's chump. When his father died, he realized he was spending enormous energy on the vigilance required to make sure he was being given his due in every situation. Even driving was draining, as he devoted his attention to whether others were cutting him off and reacted when they inevitably did. He could understand why his father wanted to pass on this value to him, his way of ensuring his son would not lose out in life. But after his father's death, he realized this was not his own value, and it was not the way he wanted to spend his mental—and physical—energy.

Another woman discovered after her father died that she was devoting her energy to dragging herself to a job every day that she had never wanted to do and got no joy from. Her father had wanted her to become a lawyer, and she had wanted to prove to him that she was capable and to make him proud. But she had never wanted to be a lawyer for any intrinsic reason. Her own motivation was squelched under layers of obligation, first to her father's dream and then to her family, who enjoyed her salary. But squelching her motivation took energy, and in addition, she did not have the restorative benefit of feeling her work was important. Her health suffered, and she had always been underweight. Grieving was hard for her, the absence of the genuine closeness of her relationship with her father, but the freedom to do something different with her life emerged. She became a yoga instructor, and her health improved along with her energy and sense of purpose.

For myself, without a strong sense that I was okay just as I was (even alone), I was constantly pulled in one direction or another, pinballing through my life. By learning how my mind worked, I was able to feel more balanced. Don't get me wrong—some days I am better at this than others. But I am more aware when my energy is being frittered away on thoughts that do not help me, on inner dialogue that prevents my awareness of where I am right now.

I learned through many months of daily meditation after the Buddhist Society Summer School that there are three types of thoughts that I find difficult to let go. They are (1) when I feel there has been a rupture in a close relationship, (2) when I feel moral outrage about something that has happened, and (3) when I have what seems like an amazing new idea. This is a great example of how even something we might think of as a good thing (eureka! discovery!) can still draw from our energy resources. Grieving people may find the would've/should've/could've thoughts are the sticky ones that are difficult to let go. The skill to recognize and choose whether the time is right to pursue a train of thought creates a greater ease in one's mind, a greater balance in mind and body as well as in the connection between them.

Connecting Mind and Body Through Awareness

Science has long divided up the mind and the body, but interrogating our conscious experience can reveal that this is not an accurate dissection, and learning to train our awareness can help us to make the best use of the interconnectedness. Here is an example: For my sister's fiftieth birthday, she invited me to go on a sailing trip with her. I was delighted, although I had never sailed on the ocean. The first couple of days went well, but then on a particularly choppy day, another passenger and I became violently sick. We took to the deck, leaning over the side to vomit, an unfamiliar experience for me since I have historically had an iron stomach. I remembered that I had been told to look at the horizon if this happened, to enable my visual and vestibular systems to find a point of common experience. Remarkably, this worked. If I kept my attention firmly trained on the horizon, my stomach settled. If I allowed my attention to wander, my stomach let me know right away. I sat on the deck like this for several hours, using intense awareness to regulate my nausea. Of course, I could not consciously experience my vestibular system; I only know that it was connected to my visual system through my gut's reaction. Importantly, none of this relied on my thoughts, just on training my focus. My body was simply reacting to the world around it, and I was able to influence this natural process by using the power of my attention.

In a similar way, grief is a physical experience, not just an emotional one. We do not choose to have grief arise or know which days will feel choppy. And verbal processing is not the only way to respond to the grief that arises in body and mind. Allowing ourselves to take the time to find balance in our mind can help us through these moments when we find our gut is upside down or our throat is in a tight knot. That pause may be breathing deeply and allowing the sensations simply to be. These sensations have information, and they will not be wished away, although learning to relax our body may help them to go away on their own. Our body is trying to understand how to be in the world without this person we love, who helped to regulate our body, and we need time, experience, and attention to do so.

In fact, our verbal focus can sometimes make painful experiences worse. Another example I can offer comes from coping with the twinkling pains that I had in my arms early in my experience of MS. After I was diagnosed, the pains continued for years. At first, they were like alarm bells for me and would capture my attention no matter what else I was doing. *Oh no! I'm having those pains again!* But over time, I realized I needed to change the way I talked to myself about the sensations. So I would have the alarm reaction and then say to myself, "You know what these are. They are not going to hurt you. They are not a sign that anything is wrong, except for a diagnosis you already know about." I would turn my attention away from them, back to the many other things that were available to my awareness, like my work or my conversation. Over many months, I was able to decrease my alarm response to these twinkling sensations, so that I would note their presence like you might notice a radio playing in the background. That is not to say that I was denying that they were happening or distracting myself. They no longer grabbed my attention, even though I was aware of them, because I no longer gave them a focus they did not warrant. They no longer prevented me from continuing to do the things that were important to me in the moment. Nothing changed about the sensations themselves, only my reaction to them.

Waves of grief can be similar—noted, accepted, and also not so disruptive as grieving changes over time. It is unrealistic to believe that at some point in the future, after we reach a state we call "closure," grief will no longer well up in us. Grief is simply the natural human emotion that arises when we become aware that our beloved is missing. As a human emotion, it is no more likely to go away than anger or sadness or disgust. But how we respond to that evoked grief can change. We can say to ourselves, "I know what this is. A wave of grief is not going to hurt me. It is not a sign that there is something wrong with me, only that my loved one is not with me now." With practice, we may be able to register that there is grief in the background of our experience, like a radio playing in the background, or a scent in the air, while we continue to pursue the many other things available to us in the present moment.

Intention, or How to Use Our Awareness and Attention

Developing awareness of our mental state enables more skillful and intentional choices. When my mind is completely focused on a task, like conceiving a lecture or answering the email of a colleague, I am mentally somewhere other than my current surroundings and my body. This requires a great deal of energy, to project myself into a different moment in time or a different place than the present one. What I have learned is that staying in that concentrated mental state for too long, productive as it might be, is not healthy for my body overall. This realization is quite a change for me from the idea that work is the most important and meaningful thing I can do. Any kind of work, even meaningful work, still needs to be balanced with rest and relaxation.

Yet when I realize I have been working for a long period and try to shift out of this intense concentration, I often find it difficult to do so. I have learned to allow myself to be aware and accept that it is difficult to shift my mental state. I continue to allow the floodlight of broad attention to the world around me to become more and more a part of my awareness, allowing the tightly focused spotlight of concentration to drop away. Sometimes it helps to stop *trying* to shift my mental state for a few minutes, especially if what I have been doing has been especially important to me, like giving a talk. Trying requires energy, and I am trying to relax into the present moment. I just notice again and again how tight my focus is, how I am running through again and again in my thoughts what just happened. No judgment, just noticing. After a period of noticing this, I will begin to shift my attention more broadly once more, to the field of possible sights and sounds and scents in my environment, and somehow it is often easier a second (or third) time.

I have also learned a few tricks that seem to help if I am having trouble relaxing my body, which usually makes it hard for me to let my thoughts go. For example, I can drop my tongue from pressing against the roof of my mouth. I put the tip of my tongue behind my top front teeth and allow the rest of my mouth and throat to relax. When my breath is rapid or shallow, I change to exhaling in the way you would breathe out to fog up your glasses to clean them, called *ujjayi*, or "ocean breath," in yoga. I find this can even help my heart rate to decelerate, capitalizing on vagal cardiac control through respiration. When I am talking with someone and I notice that I am keyed up, I think about dropping my voice. Speaking in a lower register relaxes and opens up my throat.

Setting Our Intention for When We Need It Most

Knowing when I am most likely to be tired or frazzled, when my energy will be low, or I will feel lonely, I can set an intention to react differently. These feelings often come at predictable times or regular intervals. For example, when I leave a work meeting, I can take some time to walk and breathe before I jump into the next task. It can help to shift my brooding mental state to go outside, move my body through stretching, or walk up some stairs to get my heart rate going. First thing in the morning, I can enjoy the flavor and ritual of my coffee, before I unlock my phone. You may come to discover the moments when you are most likely to feel waves of grief. Offering a little time and space to calm our grief-drenched body does wonders for our overall health. Learning to become aware of our knee-jerk reactions to low mood or low energy, and remembering to shift our reaction, does not come easily but will benefit us for the rest of our lives.

I continue to learn how to manage my mind when I am tired or feel foggy, when I find it difficult to direct my attention and let go of thoughts. Sometimes I intentionally use distraction, allow-ing my attention to be captured by something outside of my control, like binge-watching TV, or even dulling my attention through having a drink or playing solitaire on my laptop. I know this is a choice, and perhaps it would be better for me to simply rest in the present moment. In fact, one reason guided meditation is effective is that it has the added benefit of continuously helping to externally redirect one's attention. Like with Dry January, I try to reassess regularly whether I could choose other options than distraction. But I know distraction can be effective for me, like in the late evening. I constrain the length of time I spend in this state, just like I constrain the amount of time I spend concentrating on work. Distraction proves to be useful when my restless mind cannot stay in the present moment, when the attraction of guilt or worry is just too sticky. Allowing someone or something else to direct my attention is a better option for my mood than allowing my thoughts to spin. I do try to choose distractions that are not too bad for my body (more than one cocktail) or for my finances (online shopping).

At night, when I awaken and if I cannot go back to sleep, I often find my mind is gnawing away at its thoughts. Ironically, sometimes waking up a little more allows me to recall my intention to let my thoughts go. By waking up more (often by getting up out of bed), by focusing on my breathing and finding balance in my mind, I become more able to shift into sleep in a way I cannot when my attention is not under my control because I am groggy. When I return to my bed, I am more mindful of its warmth and softness, rather than whatever was previously running through my mind. I am aware of my intention to sleep, and in fact, I often say "sleep" or "ease" with each exhale. The important thing is that I have learned I can become aware of not only the content of my thoughts but also the quality of my attention and awareness. Through exploring these different mind states, I try shifting them to help me in any given situation.

Rediscovering Purpose Cannot Be Forced

The loss of a loved one can drain us with its waves of grief, the eight-foot walls of water that knock us over. We must learn to re-regulate our body and mind, to heal from being knocked off our feet by the oceanic power of our emotions. Harnessing our aware-ness and attention, and setting the intention to do so, comes with time and practice. But grieving is not only about loss-related heal-ing. It is also about restoration of a meaningful life. Rediscovering purpose takes time and effort and sometimes encouragement or support from others.

I want to be clear: When we are in acute grief, when our deep and disruptive loss is recent, we do not need to pressure ourselves to find meaning or rediscover our purpose. Simply getting through the day, or taking care of the responsibilities we still have to those around us, is difficult enough. Just show up, as who you are right now, and do the best you can. And if you are supporting some-one who is grieving, recognizing them for just showing up can be incredibly appreciated. When we are trying so hard, being told that we are doing really well in an impossible situation is deeply reassuring.

Our purpose is tied to our identity and to what we do. Psycho-logists Benjamin Bellet and Richard McNally at Harvard University were curious about how our sense of self was changed during grieving, and how this related to our activities. One of the signs of prolonged grief disorder is feeling confused about who we are, or having the sense that a part of us has died with our loved one, so much so that it continues to interrupt our day-to-day life for years after our loss. These researchers asked bereaved people to come into their lab and describe themselves by completing twenty statements that began with "I am . . ." For example, someone might write, "I am a mom who likes to cook." The researchers divided the characteristics into different categories, including demographics, physical attributes, likes and dislikes, and activities. Then they calculated the total number of unique characteristics each participant used to describe themself. Those who used fewer characteristics to describe their sense of self had more grief. Having a more variegated sense of who we are, and the roles we occupy, can be a resource when one of our roles (with our loved one) has been taken away from us. Specifically, the researchers discovered that those with fewer meaningful descriptions of their activities suffered from more severe levels of grief.[2]

If our activities (or what we do) are bound up with our loved one, such as making dinner with our spouse or caregiving for a frail parent or pet, their absence leaves a hole in what we do each day. In the wake of loss, we might not replace those activities with life-affirming ones. This leaves us with simply doing less or not feeling a clear purpose in the things we do. So not only do we lose a part of our identity, but we may also lose the agenda for our day-to-day activities. Without really knowing it, we define ourselves by what we do with our time. Because so much of what we do with our time is tied up with those people we love, we may find we do not really know how to use our time in their absence, which can be a downward spiral. We have to consciously choose new activities that are important to us, and develop these into new habits.

Bellet and McNally also discovered that bereaved people who described fewer likes and dislikes had higher levels of grief severity. Clarifying what you like, without your loved one's input, is another task of grieving. My father used to visit me from our rural hometown during the holidays, and so we would often go shopping for things he could not get at home. After my mother died, I asked what he was looking for on our next shopping trip, and he told me he needed bath towels. I took him to a department store and picked out some yellow towels to show him. "I don't want yellow towels," he told me adamantly. "I want purple towels."

"That's crazy!" I responded. "You don't want purple towels. Purple towels won't match the bathroom."

"No, I want purple towels," he said firmly.

Fortunately, I realized that my father was sharing with me an expression of who he was separate from the person he had been for the decades he was married to my mother. He had always hated yellow, apparently, but it had not mattered enough to him while she was alive to make a fuss about it. Now that she was gone, he wanted to repaint the bathroom, a reflection of his preferences, of who he was. This development of one's self as a separate person is a part of the grieving process, a reflection of one's meaningful life now. This can be painful, facing the reality of how things are now, but can also bring a sense of deep authenticity and purpose.

Rediscovering Our Purpose Requires Attention

We can benefit from rediscovering our purpose just as we have from managing our waves of grief. We must attend to how to live a more meaningful life, built for who we are now, carrying the absence of a loved one. Grieving sometimes gives us the gift of just not giving a damn. Ironically, this lack of caring about what others think may enable us to discover those activities we do find meaningful. When we pay attention to what we are doing despite our grief, we may discover those activities that sustain us. Purpose is often related to either mastery or connection, and that connection may include compassion for ourselves and others, which we will cover more in the next chapter.

Many of us find during grieving that we are not able to get going, that mastery or accomplishment feels beyond our grasp. We just cannot get invested in a hobby or our work. This demotivation can arise from two different causes, and it may be helpful to contemplate which one we are coping with. The first reason may be the fatalism that can accompany awareness of our own mortality, a hopeless feeling that nothing we do matters in the end. The second reason can be that without our loved one, we feel less motivated to explore our world.

The first reason is illustrated by an example from my own life. One summer during high school, I attended a weeklong gathering of Quakers. It was a magical experience for me, set in the stunning mountains of Montana in which I was surrounded by people who were deeply kind and thoughtful. I felt so independent and grown up, having gone to the retreat with a Quaker family I was close with, away from my mother and from the Catholicism of my childhood. Among the other young people, I met two brothers who were unlike any teenagers I had known. They had grown up in a family who nurtured their creative and emotional sides. I witnessed such genuine affection between the younger one and his mother, including him braiding her hair one night as we sat around the campfire. Their mother, Abbie, seemed to be all the things I aspired to be someday.[3]

Two weeks after the gathering, Abbie was killed by a car in a cycling accident. Although I had known her for only a brief time, her death affected me deeply. I went to her memorial, stunned by the incomprehensible fact that she could be so alive and then die suddenly, leaving her two sons motherless. I returned to my senior year of high school altered. I dropped out of a music competition that I had been preparing to do for months. I struggled to practice, despite my looming college music auditions. Practicing music had been a feature of my every day for years, and suddenly I could not bring myself to do it. What was the point? Why aspire to anything if life could be snuffed out so randomly, so suddenly?

Demotivation can come from the sudden and brutal knowledge that our life is finite and fleeting, especially when this knowledge first dawns on us. This awareness can make it seem useless to engage in a long-term project or relationship. For me and for most bereaved people, the process of taking some time away from the world, and also finding support to eventually get out there and do things again (even when we do not feel like it or see the point), leads to the slow natural upward spiral of regaining our reward motivation. I never entered a music competition again, but with reflection, I realized that going to college was still important to me, and I knew that a music scholarship was the way I could get there. I gradually got back to practicing my clarinet and did indeed get into a music school (although eventually I became a psychology major). But I will never forget how Abbie's death completely shifted my day-to-day willingness to do the things that had seemed like a given in my life.

Exploration Inhibition

There is a second reason why we can lose our motivation to do things, and this has to do with our attachment neurobiology. One of the functions of our attachment relationships is to provide us with comfort and nurturing when we return to our one-and-only, and to offer that care to them in return. This closeness with another is an imperative for our body and mind, especially when we are young or ill or sad. But that is not the only function of attachment. The other function is to enable exploration. Because we have this person to return to (our *secure base*), and because we carry them as an internal avatar as we navigate through the world, attachment enables us to explore and engage in the world. Because we have a one-and-only, a person we trust and who believes we are special, we have a sense of competence as we go about our daily activities. We know they are back there somewhere, rooting for us, cheering us on, even if only in our mind's eye. This sense of competence enables mastery as we tackle new tasks, bravery as we explore new domains. Many people describe their work as meaningful because they are doing it not just for themselves but for their family as well.

To provide, to make, to gather for our loved ones is one of the most engaging things we do.

When that loved one dies, we find ourselves without that exploration motivation. This is different from fatalistic awareness of our own mortality, like Abbie's death revealed to me, but is the lack of motivation that comes from the absence of this specific person who believed in us. Our attachment neurobiology, bereft of our attachment figure, can inhibit our exploration, seemingly draining the meaning from our activities. Avoidance of activities can go on for years after a person has lost a loved one, and can be one of the signs of prolonged grief disorder. In a type of psychotherapy designed specifically for prolonged grief, and validated in several randomized clinical trials, psychiatrist Katherine Shear at Columbia University addressed this lack of exploration as a barrier in grieving. Although prolonged grief treatment (PGT) teaches a number of skills and enables a variety of experiential exercises to facilitate new learning, this particular part of PGT focuses on fostering activities related to intrinsic interests and core values.

To address this obstacle to exploration, therapists work with grieving clients to identify a big, long-term project that could be interesting, meaningful, or satisfying. Identifying such a project can take several weeks and intentional contemplation. Skillful conversation can be required to overcome objections and inertia, then to begin creating a plan and moving forward with implementing it, for a project that will outlast the sixteen-week therapy. It does not have to be as grandiose as building the Taj Mahal (although memorialization was exactly the reason for the incredible mausoleum Shah Jahan built, his gift of architecture to the world). The project may be related to the person you were before your relationship with the deceased, or it may develop out of the life and values you shared with them. Importantly, it must be something that pulls you into your life with its gravity, pulling you toward the purpose you discover you have now.

Victor Strecher, a behavioral scientist at the University of Michigan, wrote *Life on Purpose: How Living for What Matters Changes Everything* after the death of his nineteen-year-old daughter. He describes a powerful experience, sitting in a kayak as the sun rose on Lake Michigan on Father's Day and hearing his daughter tell him that he needed to live for what mattered most. This event rocked him to the core. He went from distracting himself from his grief with mindless television to realizing he needed to go back to teaching, to teach every college student with the same purpose that he would have taught his daughter. Purpose can give us back our energy.

Awakening to What Is True Now

Finding peace and balance by learning how to use our mind can be healing, if we are willing to spend time alone in our mind, accepting that loneliness and yearning will arise, but that they will also recede. Practicing mindfulness meditation does not mean I no longer dread spending time alone. But I recognize that the dread is just a feeling, a thought, and although I do not like it, it will not hurt me, the same way that the sensations in my arms will not hurt me. Present-moment awareness is about gaining broader floodlight perspective, not only about concentration. Although guided meditation, as my mother practiced, is a good stepping stone to developing present-moment awareness, mindfulness meditation is not just about listening to someone else guide your thoughts. Meditation has given me the distance from my mind to observe myself, and the skills to calm myself and develop equanimity and compassion (including self-compassion—but more about that in the next chapter).

Spending time in silence, alone in my mind, gives me the opportunity to hear what is emerging from within me, and it helps me to establish who I am, what I think, and what I want to do with my time that aligns with my values. What would I do if no one else was weighing in, if no one else was watching? Because we carry other people in our mind, we can continue to react to what we imagine our loved ones would think, long after they have died and are no longer commenting on our activities. If these continuing bonds are comforting and make you a better person, then by all means you should conjure and commune with them. But knowing where your loved one ends and you start can be an opportunity to respond to the world, your new world, differently. Awakening to what is true now, a calm and purposeful entanglement in your world, can be energizing and a gift of grieving.

CHAPTER 10

Love and Social Connection

No one tells us that grief is going to physically hurt. The intensity of the experience of loss is hard to describe, and we tend not to hear about it from those who are in its grip. The pain of acute grief—I hate to say it, but it is true—is normal. The fact that it hurts so much is the reality of what loss feels like when someone so important to us is gone. There is nothing wrong with us. Because grief is painful, it makes sense that we might try to avoid anything that reminds us of our loss, or try to suppress the upwelling of our feelings, or find a way to go numb by throwing ourselves into work or some other activity.

Grieving may feel like we are thrashing around, just reacting to our painful experience. In yoga, I have learned that I cannot use frantic movement to keep my balance, but I can learn to find the pose that allows me to stay upright, with a certainty that I will not fall over. This place of balance and certainty can be found in the mind as well, at least for this moment. And then this moment. And then this moment. There is a place in our mind where we are balanced in our awareness of here, now, and close with everything in the world, flexibly shifting when needed between the spotlight of focus and the floodlight of broad awareness. If whatever activity we throw ourselves into is not by choice but is reactive, avoiding whatever we are thinking and feeling, then day after day, we may be spending our time and energy on things that are not aligned with our purpose and values.

But not fleeing our unpleasant feelings is not the same as brooding. Not fleeing our feelings is not the same as allowing their emotional undertow to drown us. Imagine a young woman whose boyfriend has died. For a host of reasons, she wakes up one morning feeling hopeless and depleted of energy, full of grief. What can she do? Perhaps she recognizes that she needs to engage in the world around her to put her hopeless mind in context. This is not the same as ignoring what is going on in her mind; she is not trying to distract herself from how she is feeling. But in addition to the hopeless thoughts running through her mind, she broadens her awareness to feel how cold the mug is when she takes it out of the cupboard, and then how warm it is when she takes it out of the microwave, full of hot water, ready for tea. She sits on her couch to drink it, noticing the laundry that sits unfolded next to her.

She picks up two of her white cotton socks, matching them into a pair. She is aware of how similar this is to the many times she has re-created this motion with smaller cotton socks for her beloved dog, at a time when he had a wound on his foot that needed covering. Not that she focuses on those thoughts. They simply go through her mind, through her hands, as she tucks her pair of socks in on themselves. She folds a washcloth, grateful for the loops of terry cloth that feel so good when she gets home in the evening, sticky with the day's exertion. She takes a stack of washcloths into the bathroom, bright with a yellow floor tile that makes the room seem to glow in the morning. This awareness of the moment-to-moment shift in what she is doing and the world around her, and of the accompanying thoughts that arise and recede, remind her that in just this moment, things are okay. She notices and amplifies any feelings of love and gratitude that are there, fleeting but discoverable. She may need to cope with other things when she has more mental resources, when she is not so easily pulled off-balance by how she is feeling. But for right now, tea and the couch and the laundry are all she needs to keep her balance, like looking at the horizon to settle her stomach. When the hopeless thoughts return, that's okay. In the next moment, there is laundry to fold and feel again.

Prioritizing our peace of mind could be considered our highest value. The feeling of self-compassion, the extraordinary kindness we can feel toward ourselves, enables true compassion for all other sentient beings. When I recognize that I am feeling great distress or difficulty, and because I have the desire to comfort any person who feels down like this, I can enact this compassion for myself. It may sound self-indulgent to you, but if I am not first able to comfort myself, I am no good to my partner, my students, my colleagues, or even the person I buy coffee from at the local coffee shop. A friend of mine says, "You cannot pour from an empty jug." The feeling of self-compassion, of turning kindness toward oneself, is a skill, a practice we can develop and call on when needed, when we feel off-balance.

Shaving Our Own Skin

Buddhist teacher Sharon Salzberg wrote about loving-kindness meditation in her book *Real Love: The Art of Mindful Connection.*[1] This type of meditation is a structured practice. After you settle into awareness of your breath, you imagine someone who loves you wishing you health and happiness with four phrases. Then you wish yourself and others health and happiness with the same phrases (although the phrases vary slightly with different traditions).

May I be safe.
May I be happy.
May I be healthy.
May I live with ease.

It sounds unlikely, but Salzberg described how weeks of repeating this simple meditation, which seemed to be having no effect, eventually showed up in the way she spoke to herself kindly in the midst of a chaotic situation. And receiving kind wishes is important. But perhaps more important is the embodied feeling of being generous, of uncomplicated kindness we can generate in ourselves, as the sender of those wishes. Offering unconditional love is a practice that benefits the giver, and in self-compassion, we are both the giver and the receiver.

While excellent instructions about how to do loving-kindness meditation are available,[2] my desire here is to describe the whole-hearted, embodied feeling we can locate in order to express affection toward ourselves and others. When I fell in love for the first time as a teenager, I thought this feeling was because of him, stemming from this person to whom I was so powerfully attracted. Later, after decades of experience, I learned that love actually comes from me, a feeling I continued to produce as the alchemy of two lives became entwined in a deepening relationship. Because you have loved and been loved, you can call up that experience, because you are forever changed by it. This is true even if this person has since died. The feeling of love emanates from my own mind-body, and consequently, I can find that tenderness (almost) whenever I need it. It took me practice to be able to elicit it consistently.

Oddly, discovering how to evoke this in myself was a eureka moment following an episode of the television show *Ozark*. In episode four of season two, Marty Byrde (Jason Bateman) visits the bedside of Buddy Dieker (Harris Yulin), Byrde's renter and staunch advocate. Buddy is terminally ill, and his failing health means he has not been able to shave. He mentions that his stubble is itchy. Marty offers to shave his beard, and in one of the most intimate television scenes I have witnessed, he gets shaving cream, a bowl of water, and a razor, and proceeds to shave the old man's chin. Marty looks at him so intently, so kindly, with a half smile as though he is drinking in this opportunity to connect, perhaps his last opportunity. He is focused on each movement, each stroke, each whisker that he cuts.

This scene really stuck with me. The following morning, I was still thinking about it, such a concrete demonstration of compassion.

As I got into the shower and I picked up my razor, I thought, *What if I were to shave my own leg with the same love and attention?* I looked at my leg and paid attention to what I was doing and, most importantly, allowed in myself the emergence of that same love and kindness toward my own self, through my own movements. The idea of a self-compassion meditation took on a whole new dimension, embodying the care one would give an infant or an old man, the wholehearted feeling of gratitude for this opportunity to offer kindness to one's own body, one's own self. Loving-kindness meditation is not about the words, or solely about concentration; it is about the feeling for the other person, or for one's self. There is no doubt in my mind that the change in my attitude and my demeanor resonated in the hormones and nervous system of my body.

Grief Can Keep Us from Seeing the Whole Picture

Part of the difficulty in grieving is that we are zoomed in, focused on a narrow part of our experience, on what we are thinking and feeling. Or we are focused only on what we think others want from us. Or we are focused only on the life that our loved one might have lived if they had not died. But reality has so much more breadth than this, and each moment has an infinite number of possibilities of what to notice, how to interpret our experience, and what to do.

Each of us must navigate where our own blind spots are. Our blind spots are hard to see—this is why we call them blind spots! But opening ourselves up to additional experiences, feelings, and activities can be a great place to start. My go-to strategy was helping others in order to pull myself out of being stuck in my thoughts, probably the way my mother did when my sister and I were babies. When I focused on the needs of others, I could feel better about myself, and proud of my kindness. But this functioned as a blind spot as well. Reacting to others' needs did not enable me to discover how to soothe myself, or what my own dreams and desires were. During the COVID pandemic, I had a revelation brought about by broadening my perspective—by sitting in Zoom meeting after Zoom meeting. Unlike many people who dreaded these calls, I developed a fascination with seeing myself in meetings, side by side with colleagues, noticing when I spoke up how animated I was, in direct comparison to watching others in real time.

It dawned on me that I was in a box of an equal size on the screen as each of them, no more, no less. It helped me to gain a bird's-eye perspective on situations, to include myself (and my own perspective, thoughts, and needs) as one of the group, not more or less important than each other person. And the best way forward in any situation was the way that most benefited the whole group. That is, the greatest benefit was not necessarily what I could do for others. Rather, by shifting my perspective so I saw us all with equanimity, I could consider what we all needed to do as a group. I had special insider knowledge about my own experience, of course, knowledge of my own resources and needs. But I could evaluate my contribution side by side with what they could also do and then evaluate the overall impact.

Here's a simple example: If I am running a meeting, and I am not willing to ask someone to take notes, because I do not want to impose on them, then I have to take notes while also running the meeting. It means that as a group we suffer, since my doing both things reduces the accuracy of the notes and makes the meeting move more slowly. This has made me realize that asking someone else to take notes is actually the kindest choice *from the perspective of the group*, of which I am only one part. I believe this has helped me to gain a more equanimous perspective on situations, has helped me to include myself (and my own perspective, thoughts, and needs) as one of the group, not a zero-sum game of what I can offer or they can offer.

Sharing Our Grief Can Be a Kindness

Another place I see grieving people get stuck in a narrow focus is in deciding whether to share how they are feeling. We may think we are burdening others by sharing the depth of our grief, or the fact that no, we are not over it. But as a group, we are less informed if we do not know what grief is really like, or how long it continues to affect us. The example that comes to mind is a day when my group of close friends was sitting in a restaurant, watching football and chatting and eating. Two of my friends' mothers had died within a month of each other, and the conversation turned to sharing how those friends were doing. The teenage son of one of my bereaved friends rolled his eyes at one point and said, "Do we have to talk about this? It's so depressing." My dear friend, his mother, so wisely said to him, "You know, it helps us to talk about it. I think it's okay if we stick with this topic for a few more minutes." This moment has a profound lesson for all of us, I think. As a group, we all benefited from their experience, from their willingness to help us understand our own grief, past or even future, by sharing theirs. So if we are thinking from the perspective of our whole family, or our group of friends, or even our whole culture, then the kindest thing may be to explain how it feels to be struggling with grief. This means that others come to understand that a struggle is normal after loss, and over time, they see that this struggle changes as we restore our lives.

Guilt can also result from a lack of perspective, a narrow focus. Short of premeditated murder, it was not your intentional actions that led to your loved one's death. No matter what you may *worry* you contributed—not noticing a symptom sooner, not answering their phone call, being involved in an accident with them—these are just one part of the myriad events that resulted in your loved one's death. Why do we focus only on the last day, or last weeks, of their life, rather than the hundreds of days before when we did everything in our power to demonstrate our love for them, to care for them? Why do we focus on the one time we left them without saying goodbye, instead of the many times in our relationship when we only left them after expressing our love? It is a lack of context, this narrow framing in our mind, that results in our overwhelming feelings of guilt. The memories we decide to bring to mind, the feelings of love and self-compassion that we elicit in our bodies, can be something we choose, even if memories of our darkest moments also come unbidden at times, as they will.

A shift in perspective requires being less entangled in our own story of what happened, of what we believe was cause and effect. Shifting our perspective, zooming out in terms of the number of people we are considering, or the length of time we are bracketing, can give us a new lens on the same thoughts we have been going over and over. We can see our own needs in the context of a group, see our grieving as a chapter in our lives rather than a final destination, and see sharing our grief as contributing to the ongoing knowledge of our culture. These are all shifts from first person to third person, from close-up viewfinder to wide-angle lens. I believe that this can mean seeing things just as they are, no more and no less. Forgiveness, love, and kindness toward ourselves is often about including the bigger context, just as guilt is often about leaving out aspects of the situation in order to assign blame. To see things just as they are is to forgive ourselves. To forgive ourselves allows us to see things just as they are.

To Have and To Hold

What does it mean to *have* a grieving body? Have you wished there was an owner's manual for your grieving body, for what to do with your physical pain? To have a grieving body means to stand outside yourself and consider that you have a grieving heart, a grieving nervous system. Given this fact, perhaps you can take a perspective that gives you ideas about how to take care of yourself, your health, as you move forward.

And what does it mean to *be* a grieving body? Notice how different that sounds. To be a grieving body gives you a sense of embodiment, accepting that grieving is the reality you inhabit (for now). A grieving body perceives others and events through *being* this way, while being wound up or exhausted. This acceptance of being a grieving body may lead to a different way forward, to an understanding and compassion toward your reactions to the world around you that can feel so foreign otherwise.

And what does it mean to *hold* a grieving body? To hold your grieving body, to feel the tenderness you would have for a motherless child, leads to empathic, inward-directed feelings. You might comfort yourself moving forward, giving yourself more time and attention. Each of these—to have a grieving body, to be a grieving body, to hold a grieving body—are simply changes in perspective, but this changed perspective completely alters the way forward. In the midst of grief, you nonetheless have choices of perspective and movement in response.

On Behalf of All Sentient Beings

Most Westerners think of meditation primarily for bringing about relaxation of the body. But this is not the only goal in Buddhism. Meditation is about developing wisdom and compassion. Mindfulness can also have the effect of embodied peacefulness, which is a wonderful benefit. But a loose interpretation of the phrase that Buddhists use at the end of sitting meditation is "May I use what I have learned during this meditation to benefit all sentient beings." I was mystified by this idea for a long time. How could my sitting on a cushion and focusing on my breath benefit anyone else, let alone all sentient beings? I have come to learn that meditation is just one part of cultivating greater compassion and wisdom in the Buddhist tradition. Being in the present moment does have enormous benefits for those around us, and when compassion is the impetus for why we are spending time in the present moment, it changes us as well.

In the traditional loving-kindness meditation, I have supplemented two phrases, because these phrases are the best reminders for me about how to be compassionate to myself:

> *May I be safe.*
> *May I be kind to myself.*
> *May I accept myself just the way I am.*
> *May I live with ease.*

And I offer this to others during another part of the meditation practice:

> *May you be safe.*
> *May you be kind to yourself.*
> *May you accept yourself just the way you are.*
> *May you live with ease.*

It does not matter if I can accept those around me just the way they are (although that is helpful). What matters is if they can be kind to themselves and accept themselves just the way they are.

In relationships, I try to keep that in mind. I can accept my friend or loved one just as they are, and that is an important part of my relationship with them. But this functions more as an example of how they might come to accept themselves. It means that I cannot solve what is disturbing their peace of mind, because they are the only ones who can develop their peace of mind. This way of viewing relationships feels like a great relief to me.

In many ways, I am different now in relationships than I was as a young adult with my mother. I no longer feel like I need to *do* something. I feel like I need to *be* something. I try to be a rock for my loved ones, not reaching out to help, nor withdrawing my full attention when I am with them. Not clinging to them, nor avoiding them. Being a rock takes far less energy once we become accustomed to it, although it does require practice and awareness. A rock can form a shelter for someone, without exerting any energy itself. I love through my ability to be present and kind and loving, not by helping per se, if seen from a wider perspective.

I wrote in chapter 3 that my relationship with my mother was fraught, and managing our interactions caused me great stress, anxiety, and a feeling of helplessness. I struggled to admit that my mother's death felt like a solution for me, with all the guilt that idea brought. I could not figure out how to make us both happy. But now, with a wider perspective, I understand why I viewed my mother's death as a solution, a reprieve. Because I was young, I could see no other solution to the problem of solving her suffering and my suffering, other than moving far away from her. I thought that her physical absence through death would solve my suffering. It took me years to learn how my body responded to her distress, and I would continue to react to others as I had reacted when I was with her. It took me years to heal through learning how to respond differently to distress in myself and in others.

But these lessons in healing revealed that I was just young, that I did not know any other way to solve our relationship issues in my teens and early twenties. I can see now that if she had lived, there could have been other solutions. I could have discovered, as I grew up and developed self-awareness, that I could prioritize *our relationship* needs, rather than her needs or my needs. I might have learned to be a rock, balanced and solid in myself. Her death was not the solution. That opportunity to learn other ways of relating to her was taken away from me, because she died before I could discover other solutions. And that is so sad. But losing her did not keep me from discovering those solutions, of self-awareness and healing, even though it was not in her lifetime.

I now know deeply that she was doing the best she could, and I was doing the best I could. Now I can feel grief over her loss and also understand why I was relieved to no longer have those stressful interactions with her. I know now that it was the way my life simply unfolded, not good or bad, which is perhaps a very Buddhist perspective born from my meditation practice. My continuing bond with her throughout my life has meant that I have continued to learn and to heal, and to be blessed with having had her as my deeply loving and imperfect mother.

Love Is Not Just Felt for a Specific Person

Our capacity for loving does not die when our loved one leaves this material world, and finding the ways we can continue to genuinely share what we have learned about loving from them honors their journey and our journey together. In Erich Fromm's book *The Art of Loving*, he wrote, "Love is not primarily a relationship to a specific person; it is an *attitude*, an *orientation* of *character* which determines the relatedness of the person to the world as a whole, not toward one 'object' of love."[3] We must learn how to express love as it emerges from us, rather than as a reaction to the one we love, the person who no longer lives. Fromm suggests this is how we orient our viewpoint, toward being loving. While our deceased loved one may no longer be in their physical body, our own physical body will be stronger and healthier if we use what we have learned from loving them to love others as well.

I think that once we experience love as emanating from us, we may begin to see how love can be elicited and directed toward other individuals, but also to other beloved entities. This is a love that transcends having one person in whom to invest, a love that becomes a larger form of interrelatedness. I have fallen in love with a particular narrow canyon near Tucson, Arizona. The mountains in this area of the Southwest are called "sky islands" because they are separated from other mountains and their inhabitants by the vast inhospitable desert. I love returning to this canyon, hiking the path by the creek when it is running high with monsoon water, observing its uniquely evolved hummingbird species, deer, coatimundi, woodpeckers, and wild turkeys. The feeling that wells up in me can only be described as love, and as we believe with other bonded relationships, I feel that the canyon is waiting for me even when I am not there. It has existed before me and will exist after, connecting me with generations past and generations to come, who will descend its rugged, winding path with its wide vistas and huge outcroppings of rocks.

In this larger form of interrelatedness, we can connect with our love for the natural world, spending time in green spaces or with awe-inspiring mountains or oceans. This connection transcends our own small self and overcomes our internal focus, which research suggests may reduce our blood pressure, deepen our breathing, and improve our neurochemistry. Our body is in a relationship with the wider world, as it is in a relationship with our loved one. Our heart rate rises as we pump our muscles to soldier up the steep mountain, our melatonin levels respond to the lack of artificial light at night in uninhabited places. Our body is composed of, and in conversation with, the world around us.

On a smaller scale, when I am meditating, I focus on the here, now, and connectedness of mindfulness from this transcendent perspective. I focus my attention on the closeness between me and anything that is present in my environment: the overlap between myself and the houseplants in the room, giving and taking in the same oxygen; myself and the other people I might be meditating with, or even the people who are meditating somewhere in the world; myself and the blueness of the rug beneath my cushion, and the fact that the sun's light is falling equally on me and the rug. I have many versions of this interconnectedness feeling, and I can usually find one that feels true and alive in that moment.

I began this book by saying that investigating how we can die of a broken heart has been one of the motivating questions in my career. Why do we care that people might die of a broken heart? Of course, we do not want to die, and fear of our own mortality might be one reason to better understand it. We also do not want our loved ones to die of a broken heart, and fear of loss might be another reason. But I think there is yet another reason. The very evidence that we can die of a broken heart means that our physical body is made up of our connections with our loved ones, regulated by their presence in our life and in our mind. Mine is not a motivation to study this phenomenon that comes from fear. It comes from the fascination and wonder and awe that two separate bodies can come together through attachment in this way, tethered through invisible bonds, creating our deepest motivations and strongest emotions, transcending time and space.

Loss Can Be a Turning Point

As a scientist, I worry about telling you too much about a sample size of only one, my own idiosyncratic losses, which feels at odds with the goal of understanding the universal rules of the human grief experience. But I want you to know that although I may be a bereavement scientist, I am also a griever. I share my internal experience with you on these pages not because yours will be the same but because I have had the privilege in this life to explore what it means to have and be a grieving body, with all the available cutting-edge scientific information about how that body works. I do not mistake my personal grief experience for expertise—I cannot give you advice, because your experience will be different from mine in ways I cannot predict. But hopefully, like lending you my glasses, my view of the physiology of grief may help you to bring into focus some aspects of your own life that you could not see before.

Loss can affect our physical health, our immune and endocrine and cardiovascular functioning. Put differently, the grief we feel may be simply the consequence, the output, of our systems reregulating after our loved one is gone. Grief may not be the problem we have to deal with per se, but rather, if we restore a meaningful and healthy life, we may be able to accept that grief is only one of myriad feelings.

Although it is clear that the death of a loved one is one of life's greatest stresses, *why* we feel grief may differ. For myself, accepting that the panic of being alone was a part of the human experience, that I was going to die someday, that I could not unburden others' of their suffering, these were all a part of the *why*. My loss of health, of the energy to which I believed I was entitled, could have gotten in the way of my working as a scientist. Learning to listen to my body's stress signals improved my physical health and allowed me to heal, by which I mean to function holistically. To be fair, I had a great number of resources benefiting me in my pursuit of scientific work that not everyone has. But sitting in silence and listening to my body's experience enabled me to do better science. Science is a powerful method but does not tell us what questions to ask or what to do with the data. My experiences of loss led me to ask different questions, to value communicating grief research more widely to the public, and to build the field of grief research by integrating medicine and neuroscience and psychology. Loss involves so much disruption that it can be a turning point in developing a healthy life that is balanced, an opportunity for physical healing. If we allow the awareness of the fragility and brevity of life to change our goals and values and compassion, we can change the world toward a greater expression of love, because love requires the deep knowledge of loss.

Acknowledgments

Merriam-Webster suggests that "acknowledge" implies disclosing something that might be concealed, as well as expressing gratitude,[*] and it is my purpose here both to name many of the people who have contributed to writing this book and also express my deep gratitude to each of them. Most of all, I acknowledge and thank the many unnamed people who have shared their grief story with me, as they are the observers with whom any scientific understanding must be built.

Most directly contributing to the birth of this book, I want to thank my amazing, wise, thoughtful (and ever-responsive!) agent, Laurie Abkemeier, and everyone at DeFiore and Company. To my

---

[*] https://www.merriam-webster.com/thesaurus/acknowledge.

editor, Gabriella Page-Fort, who took on this project with gusto and turned it into a reality, and to associate editor Maya Alpert, copyeditor Dianna Stirpe, publicist Louise Braverman, and everyone working at HarperOne—thank you for the (sometimes impossible-seeming) achievement of bringing books to the world and ideas to people. It requires more parts working smoothly in concert than an author could ever truly understand.

I am grateful that the whole purpose of academia is to disseminate what is learned by researchers. I see this every day in the time professors spend with their mentees, painstakingly pointing out errors in technique or logic but also supporting and encouraging their new ideas. I am grateful to each of my graduate students, who continue to show me the world in a new way through their eyes. Many thanks to Diana Leonard, who taught me to really listen to the human voice. I am grateful to those who were willing to read portions of the book and comment on them (although any errors are my own): Mustafa Al'Absi, Steve Cole, Lauren DePino, Al Kaszniak, Lance Meeks, Lizzie Pickering, Karen Quigley, Myleigha Truitt, and Kate Wolitzky-Taylor. Most of all, to Dave Sbarra, who read every draft page and commented in ways that made the book more scientifically rigorous and more readable—my gratitude exceeds a good bottle of whiskey.

To the staff of Flora's Market Run, a true Tucson neighborhood institution, for knowing my regular coffee (or tea!) order and providing me a homey place to write every morning (and ridiculously good baked goods).

And to my loved ones—I am grateful for your emotional support, which makes it possible for me not only to write books but also to be a happy and compassionate person. For Rick Grossman,

Caroline O'Connor, Beth Peterson, and Anna Visscher, you make it all possible and worthwhile. For the gang—the Keemes, Nolens, and Petersons—for reminding me to have fun and making me laugh. To my father, who taught me how to live a meaningful life and what a good death can look like. Finally, to my mother, my unending wordless gratitude for all that you were and all that you became, and all the ways you live on in me, for the difficulties and the joys, but most of all for your unceasing love and belief in me. This is not your story, but my story is immeasurably shaped by yours.

Notes

Introduction: When a Part of "Us" Has Been Cut Away

1. Lisa Feldman Barrett and John Dunne, "Buddhists in Love," *Aeon*, aeon.co/essays/does-buddhist-detachment-allow-for-a-healthier -togetherness.
2. Lisa Feldman Barrett, *How Emotions Are Made: The Secret Life of the Brain* (Houghton Mifflin Harcourt, 2017).
3. P. Martikainen and T. Valkonen, "Mortality After the Death of a Spouse: Rates and Causes of Death in a Large Finnish Cohort," *American Journal of Public Health* 86, no. 8, pt. 1 (1996): 1087–93, doi.org /10.2105/ajph.86.8_pt_1.1087.

Chapter 1: The Heart

1. Eduardo Medina, "The Husband of a Teacher Killed in the Massacre Has Died of a Heart Attack," *New York Times*, May 26, 2022, www. nytimes.com/2022/05/26/us/irma-garcia-husband-death-uvalde.html.

2. E. Mostofsky et al., "Risk of Acute Myocardial Infarction After the Death of a Significant Person in One's Life: The Determinants of Myocardial Infarction Onset Study," *Circulation* 125, no. 3 (2012): 491–96, doi.org/10.1161/CIRCULATIONAHA.111.061770.

3. M.-F. O'Connor et al., "Emotional Disclosure for Whom? The Role of Vagal Tone in Bereavement," *Biological Psychology* 68, no. 2 (2005): 135–46.

4. C. Templin et al., "Clinical Features and Outcomes of Takotsubo (Stress) Cardiomyopathy," *New England Journal of Medicine* 373, no. 10 (2015): 929–38, doi.org/10.1056/NEJMoa1406761.

5. J. R. Ghadri et al., "Happy Heart Syndrome: Role of Positive Emotional Stress in Takotsubo Syndrome," *European Heart Journal* 37, no. 37 (2016): 2823–29, doi.org/10.1093/eurheartj/ehv757.

6. T. Buckley et al., "Physiological Correlates of Bereavement and the Impact of Bereavement Interventions," *Dialogues in Clinical Neuroscience* 14, no. 2 (2012): 129–39, doi.org/10.31887/DCNS.2012.14.2/tbuckley.

7. R. Palitsky et al., "The Relationship of Prolonged Grief Disorder Symptoms with Hemodynamic Response to Grief Recall Among Bereaved Adults," *Psychosomatic Medicine* 85, no. 6 (2023): 545–50, doi.org/10.1097/PSY.0000000000001223.

8. J. Coutinho et al., "When Our Hearts Beat Together: Cardiac Synchrony as an Entry Point to Understand Dyadic Co-regulation in Couples," *Psychophysiology* 58, no. 3 (2021): e13739, doi.org/10.1111/psyp.13739.

9. M. A. Hofer, "Relationships as Regulators: A Psychobiologic Perspective on Bereavement," *Psychosomatic Medicine* 46, no. 3 (1984): 183–97, doi.org/10.1097/00006842–198405000–00001.

10. S. Karl et al., "Low-Dose Aspirin for Prevention of Cardiovascular Risk in Bereavement: Results from a Feasibility Study," *Psychotherapy and Psychosomatics* 87 (2018): 112-113, doi.org/10.1159/000481862.

11. G. H. Tofler et al., "The Effect of Metoprolol and Aspirin on Cardiovascular Risk in Bereavement: A Randomized Controlled Trial," *American Heart Journal* 220 (2020): 264–72, doi.org/10.1016/j.ahj.2019.11.003.

Chapter 2: The Immune System

1. C. R. Schultze-Florey et al., "When Grief Makes You Sick: Bereavement Induced Systemic Inflammation Is a Question of Genotype,"

Brain, Behavior, and Immunity 26, no. 7 (2012): 1066–71, doi.org/10.1016/j.bbi.2012.06.009.

2. C. P. Fagundes et al., "Grief, Depressive Symptoms, and Inflammation in the Spousally Bereaved," *Psychoneuroendocrinology* 100 (2019): 190–97, doi.org/10.1016/j.psyneuen.2018.10.006; and C. P. Fagundes et al., "Spousal Bereavement Is Associated with More Pronounced ex Vivo Cytokine Production and Lower Heart Rate Variability: Mechanisms Underlying Cardiovascular Risk?," *Psychoneuroendocrinology* 93 (2018): 65–71, doi.org/10.1016/j.psyneuen.2018.04.010.

3. A. C. Phillips et al., "Bereavement and Marriage Are Associated with Antibody Response to Influenza Vaccination in the Elderly," *Brain, Behavior, and Immunity* 20, no. 3 (2006): 279–89, doi.org/10.1016/j.bbi.2005.08.003.

4. A. S. LeRoy et al., "Implications for Reward Processing in Differential Responses to Loss: Impacts on Attachment Hierarchy Reorganization," *Personality and Social Psychology Review* 23, no. 4 (2019): 391–405, doi.org/10.1177/1088868319853895.

5. R. Dantzer et al., "From Inflammation to Sickness and Depression: When the Immune System Subjugates the Brain," *Nature Reviews Neuroscience* 9, no. 1 (2008): 46–56, doi.org/10.1038/nrn2297.

6. M. Maes et al., "The New '5-HT' Hypothesis of Depression: Cell-Mediated Immune Activation Induces Indoleamine 2,3-Dioxygenase, Which Leads to Lower Plasma Tryptophan and an Increased Synthesis of Detrimental Tryptophan Catabolites (TRYCATs), Both of Which Contribute to the Onset of Depression," *Progress in Neuro-Psychopharmacology and Biological Psychiatry* 35, no. 3 (2011): 702–21, doi.org/10.1016/j.pnpbp.2010.12.017.

7. M. Melnikov and A. Lopatina, "Th17-Cells in Depression: Implication in Multiple Sclerosis," *Frontiers in Immunology* 13 (2022): 1010304, doi.org/10.3389/fimmu.2022.1010304.

Chapter 3: The Endocrine System

1. A. Enmalm and R. Boehme, "Body Perception and Social Touch Preferences in Times of Grief," *Journal of Loss and Trauma* (February 20, 2024): 1–24, doi.org/10.1080/15325024.2024.2316117.

2. L. Demarchi, A. Sanson, and O. J. Bosch, "Neurobiological Traces of Grief: Examining the Impact of Offspring Loss After Birth on Rat Mothers' Brain and Stress-Coping Behavior in the First Week

Postpartum," *Neuroscience Applied* 3 (2024): 104065, doi.org/10.1016/j
.nsa.2024.104065.

3. D. C. Wigger et al., "Maternal Separation Induces Long-Term Altera-
tions in the Cardiac Oxytocin Receptor and Cystathionine γ-Lyase Expression in Mice," *Oxidative Medicine and Cellular Longevity* 2020, no. 1
(2020): 4309605, doi.org/10.1155/2020/4309605.

4. T. Buckley et al., "Physiological Correlates of Bereavement and the Impact of Bereavement Interventions," *Dialogues in Clinical Neuroscience*
14, no. 2 (2012): 129–39, doi.org/10.31887/DCNS.2012.14.2/tbuckley.

5. A. J. Grippo et al., "Behavioral and Neuroendocrine Consequences of
Disrupting a Long-Term Monogamous Social Bond in Aging Prairie
Voles," *International Journal on the Biology of Stress* 24, no. 3 (2021):
239–50, doi.org/10.1080/10253890.2020.1812058.

6. M. A. Fallon et al., "Utility of a Virtual Trier Social Stress Test: Initial
Findings and Benchmarking Comparisons," *Psychosomatic Medicine*
78, no. 7 (2016): 835–40, doi.org/10.1097/PSY.0000000000000338.

7. H. C. Saavedra Pérez et al., "The Impact of Complicated Grief on
Diurnal Cortisol Levels Two Years After Loss: A Population-Based
Study," *Psychosomatic Medicine* 79, no. 4 (2017): 426–33, doi.org/10.1097
/PSY.0000000000000422.

8. E. K. Adam et al., "Diurnal Cortisol Slopes and Mental and Physical
Health Outcomes: A Systematic Review and Meta-Analysis," *Psychoneuroendocrinology* 83 (2017): 25–41, doi.org/10.1016/j.psyneuen.2017.05.018.

9. Richard S. Lazarus and Susan Folkman, *Stress, Appraisal, and Coping*
(Springer, 1984).

10. M. Stroebe and H. Schut, "The Dual Process Model of Coping with
Bereavement: A Decade On," *Omega* 61, no. 4 (2010): 273–289, doi.org
/10.2190/OM.61.4.b

11. M. Chang and T. F. Robles, "Losing Loved Ones Too Soon and Too
Much: Measuring Racial Disparities in Lifetime Loss Burden," PsyArXiv
Preprints (February 23, 2024): 1–34, doi.org/10.31234/osf.io/5tvgz.

12. T. T. Lewis et al., "Race/Ethnicity, Cumulative Midlife Loss, and
Carotid Atherosclerosis in Middle-Aged Women," *American Journal of
Epidemiology* 190, no. 4 (2021): 576–87, doi.org/10.1093/aje/kwaa213.

13. Elizabeth Arias, Betzaidad Tejada-Vera, Farida Ahmad, and Kenneth
D. Kochanek, *Vital Statistics Rapid Release*, report no. 15 (National

Center for Health Statistics, 2021), 2, www.cdc.gov/nchs/data/vsrr /vsrr015–508.pdf.

14. Breeshia Wade, *Grieving While Black: An Antiracist Take on Oppression and Sorrow* (North Atlantic Books, 2021), 26.

Chapter 4: The Liver and Lungs

1. T. Buckley et al., "Prospective Study of Early Bereavement on Psychological and Behavioural Cardiac Risk Factors," *Internal Medicine Journal* 39, no. 6 (2009): 370–78, doi.org/10.1111/j.1445–5994.2009.01879.x.

2. J. Pilling et al., "Alcohol Use in the First Three Years of Bereavement: A National Representative Survey," *Substance Abuse Treatment, Prevention, and Policy* 7, no. 1 (2012): 3, doi.org/10.1186/1747–597X-7–3.

3. A. Pitman et al., "Self-Reported Patterns of Use of Alcohol and Drugs After Suicide Bereavement and Other Sudden Losses: A Mixed Methods Study of 1,854 Young Bereaved Adults in the UK," *Frontiers in Psychology* 11 (2020): 1024, doi.org/10.3389/fpsyg.2020.01024.

4. S. G. Christiansen et al., "Alcohol-Related Mortality Following the Loss of a Child: A Register-Based Follow-Up Study from Norway," *BMJ Open* 10, no. 6 (2020): e038826, doi.org/10.1136/bmjopen-2020–038826.

5. G. A. Bonanno and C. L. Burton, "Regulatory Flexibility: An Individual Differences Perspective on Coping and Emotion Regulation," *Perspectives on Psychological Science* 8, no. 6 (2013): 591–612, doi.org/10 .1177/1745691613504116.

6. Diane Eshin Rizzetto, *Waking Up to What You Do* (Shambhala, 2005).

7. C. Sripada, "Impaired Control in Addiction Involves Cognitive Distortions and Unreliable Self-Control, Not Compulsive Desires and Overwhelmed Self-Control," *Behavioural Brain Research* 418 (2022): 113639, doi.org/10.1016/j.bbr.2021.113639.

8. J.-K. Zubieta et al., "Regulation of Human Affective Responses by Anterior Cingulate and Limbic μ-Opioid Neurotransmission," *Archives of General Psychiatry* 60, no. 11 (2003): 1145–53, doi.org/10.1001 /archpsyc.60.11.1145.

9. D. Shaw and M. al'Absi, "Attenuated Beta Endorphin Response to Acute Stress Is Associated with Smoking Relapse," *Pharmacology, Biochemistry, and Behavior* 90, no. 3 (2008): 357–62, doi.org/10.1016/j.pbb .2008.03.020.

10. T. K. Inagaki et al., "Opioids and Social Bonding: Naltrexone Reduces Feelings of Social Connection," *Social Cognitive and Affective Neuroscience* 11, no. 5 (2016): 728–35, doi.org/10.1093/scan/nsw006.

11. N. C. Christie, "The Role of Social Isolation in Opioid Addiction," *Social Cognitive and Affective Neuroscience* 16, no. 7 (2021): 645–56, doi.org/10.1093/scan/nsab029.

12. K. J. Bourassa, J. M. Ruiz, and D. A. Sbarra, "The Impact of Physical Proximity and Attachment Working Models on Cardiovascular Reactivity: Comparing Mental Activation and Romantic Partner Presence," *Psychophysiology* 56, no. 5 (2019): e13324, doi.org/10.1111/psyp.13324.

13. J. Younger et al., "Viewing Pictures of a Romantic Partner Reduces Experimental Pain: Involvement of Neural Reward Systems," *PLoS One* 5, no.10 (2010): e13309, doi.org/10.1371/journal.pone.0013309.

14. N. J. van Haeringen, "The (Neuro)anatomy of the Lacrimal System and the Biological Aspects of Crying," chap. 2 in *Adult Crying: A Biopsychosocial Approach*, ed. A. J. J. Vingerhoets and Randolph R. Cornelius (Routledge, 2001), doi.org/10.4324/9780203717493.

15. M. C. Eisma and M. S. Stroebe, "Rumination Following Bereavement: An Overview," *Bereavement Care* 36, no. 2 (2017): 58–64, doi.org/10.1080/02682621.2017.1349291.

16. M. C. Eisma et al., "Is Rumination After Bereavement Linked with Loss Avoidance? Evidence from Eye-Tracking," *PloS One* 9, no. 8 (2014): e104980, doi.org/10.1371/journal.pone.0104980.

Chapter 5: The Brain

1. L. McWhirter et al., "What Is Brain Fog?," *Journal of Neurology, Neurosurgery, & Psychiatry* 94, no. 4 (2023): 321–25, doi.org/10.1136/jnnp-2022-329683.

2. J. B. Badenoch et al., "Persistent Neuropsychiatric Symptoms After COVID-19: A Systematic Review and Meta-Analysis," *Brain Communications* 4, no. 1 (2022): fcab297, doi.org/10.1093/braincomms/fcab297.

3. C. R. Schultze-Florey et al., "When Grief Makes You Sick: Bereavement Induced Systemic Inflammation Is a Question of Genotype," *Brain,*

Behavior, and Immunity 26, no. 7 (2012): 1066–71, doi.org/10.1016/j.bbi.2012.06.009.

4. A. Hugoson, B. Ljungquist, and T. Breivik, "The Relationship of Some Negative Events and Psychological Factors to Periodontal Disease in an Adult Swedish Population 50 to 80 Years of Age," *Journal of Clinical Periodontology* 29, no. 3 (2002): 247–53, doi.org/10.1034/j.1600–051x.2002.290311.x.

5. Schultze-Florey et al., "When Grief Makes You Sick."

6. Hugoson, Ljungquist, and Breivik, "The Relationship of Some Negative Events."

7. H. Wang et al., "Microglia in Depression: An Overview of Microglia in the Pathogenesis and Treatment of Depression," *Journal of Neuroinflammation* 19, no. 1 (2022): 132, doi.org/10.1186/s12974–022–02492–0.

8. G. M. Slavich and M. R. Irwin, "From Stress to Inflammation and Major Depressive Disorder: A Social Signal Transduction Theory of Depression," *Psychological Bulletin* 140, no. 3 (2014): 774–815, doi.org/10.1037/a0035302.

9. Y. Zhan et al., "Deficient Neuron-Microglia Signaling Results in Impaired Functional Brain Connectivity and Social Behavior," *Nature Neuroscience* 17, no. 3 (2014): 400–6, doi.org/10.1038/nn.3641.

10. J. Steiner et al., "Severe Depression Is Associated with Increased Microglial Quinolinic Acid in Subregions of the Anterior Cingulate Gyrus: Evidence for an Immune-Modulated Glutamatergic Neurotransmission?," *Journal of Neuroinflammation* 8, no. 94 (2011), doi.org/10.1186/1742–2094–8–94.

11. H. Wang et al., "Microglia in Depression: An Overview of Microglia in the Pathogenesis and Treatment of Depression," *Journal of Neuroinflammation* 19, no. 1 (2022): 132, doi.org/10.1186/s12974–022–02492–0.

12. National Institute of Environmental Health Sciences, "Microbiome," last modified March 22, 2024, www.niehs.nih.gov/health/topics/science/microbiome/index.cfm.

13. M. Maes, "The Cytokine Hypothesis of Depression: Inflammation, Oxidative & Nitrosative Stress (IO&NS) and Leaky Gut as New Targets for Adjunctive Treatments in Depression," *Neuro Endocrinology Letters* 29, no. 3 (2008): 287–91, www.nel.edu/userfiles/articlesnew/NEL290308R02.pdf.

14. A. S. Carlessi et al., "Gut Microbiota–Brain Axis in Depression: The Role of Neuroinflammation," *European Journal of Neuroscience* 53, no.1 (2021): 222–35, doi.org/10.1111/ejn.14631.

15. Lauren DePino, "Grief Made Me Lose My Balance. Here's How I Learned to Walk Forward Again," NPR, March 38, 2024, www.npr .org/sections/health-shots/2024/03/28/1241316836/grief-accident-prone -loss-recovery-falls.

16. Martikainen and Valkonen, "Mortality After the Death of a Spouse."

17. S. M. Aoun et al., "Who Needs Bereavement Support? A Population Based Survey of Bereavement Risk and Support Need," *PLoS One* 10, no. 3 (2015): e0121101, doi.org/10.1371/journal.pone.0121101.

18. M. Stroebe, H. Schut, and W. Stroebe, "Health Outcomes of Bereavement," *The Lancet* 370, no. 9603 (2007): 1960–73, doi.org/10.1016/S0140 –6736(07)61816–9.

19. Suzanne Crawford O'Brien, *Religion and Healing in Native America: Pathways for Renewal* (Praeger, 2008).

20. K. M. Gothard and A. J. Fuglevand, "The Role of the Amygdala in Processing Social and Affective Touch," *Current Opinion in Behavioral Sciences* 43 (2022): 46–53, doi.org/10.1016/j.cobeha.2021.08.004; and E. Schneider et al., "Affectionate Touch and Diurnal Oxytocin Levels: An Ecological Momentary Assessment Study," *eLife* 12 (2023): e81241, doi.org/10.7554/eLife.81241.

Chapter 6: The Sympathetic Nervous System

1. J. K. Allen et al., "Sustained Adrenergic Signaling Promotes Intratumoral Innervation Through BDNF Induction," *Cancer Research* 78, no. 12 (2018): 3233–42, doi.org/10.1158/0008–5472.CAN-16–1701.

2. American Association of Neurological Surgeons (AANS) et al., "Multisociety Consensus Quality Improvement Revised Consensus Statement for Endovascular Therapy of Acute Ischemic Stroke," *American Journal of Neuroradiology* 39, no. 6 (2018): E61–76, doi.org/10.3174/ajnr .A5638.

3. Khan Academy, "Intro to Gene Expression (Central Dogma)," AP/College Biology course overview, accessed July 15, 2024, www.khanacademy.org /science/ap-biology/gene-expression-and-regulation/translation/a/intro -to-gene-expression-central-dogma.

4. S. W. Cole et al., "Social Regulation of Gene Expression in Human Leukocytes," *Genome Biology* 8, no. 9 (2007): R189, doi.org/10.1186/gb-2007-8-9-r189.

5. D. S. Black, G. Christodoulou, and S. Cole, "Mindfulness Meditation and Gene Expression: A Hypothesis-Generating Framework," *Current Opinion in Psychology* 28 (2019): 302–6, doi.org/10.1016/j.copsyc.2019.06.004.

6. M.-F. O'Connor et al., "Divergent Gene Expression Responses to Complicated Grief and Non-complicated Grief," *Brain, Behavior, and Immunity* 37 (2014): 78–83, doi.org/10.1016/j.bbi.2013.12.017.

7. F. Elwert and N. A. Christakis, "The Effect of Widowhood on Mortality by the Causes of Death of Both Spouses," *American Journal of Public Health* 98, no. 11 (2008): 2092–98, doi.org/10.2105/AJPH.2007.114348.

8. A. C. Phillips et al., "Bereavement and Marriage Are Associated with Antibody Response to Influenza Vaccination in the Elderly," *Brain, Behavior, and Immunity* 20, no. 3 (2006): 279–89, doi.org/10.1016/j.bbi.2005.08.003.

9. A. S. LeRoy et al., "Relationship Satisfaction Determines the Association Between Epstein-Barr Virus Latency and Somatic Symptoms After the Loss of a Spouse," *Personal Relationships* 27, no. 3 (2020): 652–73, doi.org/10.1111/pere.12336.

10. Allen et al., "Sustained Adrenergic Signaling Promotes Intratumoral Innervation."

11. M. H. Antoni and F. S. Dhabhar, "The Impact of Psychosocial Stress and Stress Management on Immune Responses in Patients with Cancer," *Cancer* 125, no. 9 (2019): 1417–31, doi.org/10.1002/cncr.31943.

12. L. M. Knowles et al., "A Controlled Trial of Two Mind-Body Interventions for Grief in Widows and Widowers," *Journal of Consulting and Clinical Psychology* 89, no. 7 (2021): 640–54, doi.org/10.1037/ccp0000653.

13. J. Wagner et al., "Community Health Workers Assisting Latinos Manage Stress and Diabetes (CALMS-D): Rationale, Intervention Design, Implementation, and Process Outcomes," *Translational Behavioral Medicine* 5, no. 4 (2015): 415–24, doi.org/10.1007/s13142-015-0332-1.

14. Douglas A. Bernstein, Thomas D. Borkovec, and Holly Hazlett-Stevens, *New Directions in Progressive Relaxation Training: A Guidebook for Helping Professionals* (Praeger, 2000).

15. S. Dimidjian and Z. V. Segal, "Prospects for a Clinical Science of Mindfulness-Based Intervention," *American Psychologist* 70, no. 7 (2015): 593–620, doi.org/10.1037/a0039589.

16. Y. Kim et al., "Only the Lonely: Expression of Proinflammatory Genes Through Family Cancer Caregiving Experiences," *Psychosomatic Medicine* 83, no. 2 (2021): 149–56, doi.org/10.1097/PSY.0000000000000897.

17. B. L. Fredrickson et al., "Psychological Well-Being and the Human Conserved Transcriptional Response to Adversity," *PLoS One* 10, no. 3 (2015): e0121839, doi.org/10.1371/journal.pone.0121839.

Chapter 7: Energy and Motivation

1. Ed Yong, "Fatigue Can Shatter a Person," *The Atlantic*, July 27, 2023, www.theatlantic.com/health/archive/2023/07/chronic-fatigue-long-covid-symptoms/674834/.

2. G. M. Cooney et al., "Exercise for Depression," *Cochrane Database of Systematic Reviews* 2013, no. 9 (2013): CD004366, doi.org/10.1002/14651858 .CD004366.pub6; and M. Noetel et al., "Effect of Exercise for Depression: Systematic Review and Network Meta-Analysis of Randomised Controlled Trials," *BMJ (Clinical Research Edition)* 384 (2024): e075847, doi.org/10.1136/bmj-2023–075847.

3. Yong, "Fatigue Can Shatter a Person."

4. Global Wind Energy Council, "Examining the 'Second Shift' for Working Women," May 13, 2019, gwec.net/examining-the-second-shift -for-working-women/; and Emily Field, Alexis Krivkovich, Sandra Kügele, Nicole Robinson, and Lareina Yee, *Women in the Workplace 2023* (McKinsey & Company, October 5, 2023), www.mckinsey.com /featured-insights/diversity-and-inclusion/women-in-the-workplace.

5. Yong, "Fatigue Can Shatter a Person."

6. Megan Devine, *It's OK That You're Not OK: Meeting Grief and Loss in a Culture That Doesn't Understand* (Sounds True, 2017).

Chapter 8: Healthy Habits

1. D. Hopf et al., "Loneliness and Diurnal Cortisol Levels During COVID-19 Lockdown: The Roles of Living Situation, Relationship Status, and Relationship Quality," *Scientific Reports* 12, no. 1 (2022): 15076, doi.org/10.1038/s41598–022–19224–2.

2. D. A. Chirinos (January 29, 2022). "Sleep Disturbances in Spousal Bereavement: Developing a Targeted Intervention," Neurobiology of Grief International Network (NOGIN) Conference, Tucson, AZ.

3. J. Warner, C. Metcalfe, and M. King, "Evaluating the Use of Benzodiazepines Following Recent Bereavement," *British Journal of Psychiatry* 178, no. 1 (2001): 36–41, doi.org/10.1192/bjp.178.1.36.

4. A. J. Spielman, L. S. Caruso, and P. B. Glovinsky, "A Behavioral Perspective on Insomnia Treatment," *Psychiatric Clinics of North America* 10, no. 4 (1987): 541–53, doi.org/10.1016/S0193–953X(18)30532-X.

5. M. Lancel, M. Stroebe, and M. C. Eisma, "Sleep Disturbances in Bereavement: A Systematic Review," *Sleep Medicine Reviews* 53 (2020): 101331, doi.org/10.1016/j.smrv.2020.101331.

6. E. E. Beem et al., "Psychological Functioning of Recently Bereaved, Middle-Aged Women: The First 13 Months," *Psychological Reports* 87, no. 1 (2000): 243–54, doi.org/10.2466/pr0.2000.87.1.243.

7. Society of Behavioral Sleep Medicine, "United States" (member map), accessed July 15, 2024, www.behavioralsleep.org/index.php/united-states-sbsm-members.

8. M. de Feijter et al., "The Longitudinal Association of Actigraphy-Estimated Sleep with Grief in Middle-Aged and Elderly Persons," *Journal of Psychiatric Research* 137 (2021): 66–72, doi.org/10.1016/j.jpsychires.2021.02.042.

9. A. Germain et al., "Treating Complicated Grief: Effects on Sleep Quality," *Behavioral Sleep Medicine* 4, no. 3 (2006): 152–63, doi.org/10.1207/s15402010bsm0403_2; and P. A. Boelen and J. Lancee, "Sleep Difficulties Are Correlated with Emotional Problems Following Loss and Residual Symptoms of Effective Prolonged Grief Disorder Treatment," *Depression Research and Treatment* (2013): 739804, doi.org/10.1155/2013/739804.

10. P. A. Carter, S. Q. Mikan, and C. Simpson, "A Feasibility Study of a Two-Session Home-Based Cognitive Behavioral Therapy–Insomnia Intervention for Bereaved Family Caregivers," *Palliative & Supportive Care* 7, no. 2 (2009): 197–206, doi.org/10.1017/S147895150900025X.

11. J. E. Carroll et al., "Sleep and Multisystem Biological Risk: A Population-Based Study," *PLoS One* 10, no. 2 (2015): e0118467, doi.org/10.1371/journal.pone.0118467.

12. C. A. Rosenbloom and F. J. Whittington, "The Effects of Bereavement on Eating Behaviors and Nutrient Intakes in Elderly Widowed Persons," *Journal of Gerontology* 48, no. 4 (1993): S223–29, doi.org/10.1093/geronj/48.4.s223.

13. S. Lee et al., "Effects of Marital Transitions on Changes in Dietary and Other Health Behaviours in US Women," *International Journal of Epidemiology* 34, no. 1 (2005): 69–78, doi.org/10.1093/ije/dyh258.

14. D. R. Shahar et al., "The Effect of Widowhood on Weight Change, Dietary Intake, and Eating Behavior in the Elderly Population," *Journal of Aging and Health* 13, no. 2 (2014): 189–99, doi.org/10.1177/089826430101300202.

15. S. Wilcox et al., "The Effects of Widowhood on Physical and Mental Health, Health Behaviors, and Health Outcomes: The Women's Health Initiative," *Health Psychology* 22, no. 5 (2003): 513–22, doi.org/10.1037/0278–6133.22.5.513.

16. E. Vesnaver et al., "Food Behavior Change in Late-Life Widowhood: A Two-Stage Process," *Appetite* 95 (2015): 399–407, doi.org/10.1016/j.appet.2015.07.027.

17. L. Tomova et al., "Author Correction: Acute Social Isolation Evokes Midbrain Craving Responses Similar to Hunger," *Nature Neuroscience* 25, no. 3 (2022): 399, doi.org/10.1038/s41593–021–01004–2.

18. Kristen Neff, Self-Compassion website, self-compassion.org/.

19. Lucy Foulkes, "How to Have More Meaningful Conversations," *Psyche*, April 28, 2021, psyche.co/guides/how-to-have-more-meaningful-conversations.

20. Irvin D. Yalom and Marilyn Yalom, *A Matter of Life and Death* (Redwood Press, 2021).

21. F. Bruno et al., "Positive Touch Deprivation During the COVID-19 Pandemic: Effects on Anxiety, Stress, and Depression Among Italian General Population," *Brain Sciences* 13, no. 4 (2023): 540, doi.org/10.3390/brainsci13040540.

22. K. C. Light, K. M. Grewen, and J. A. Amico, "More Frequent Partner Hugs and Higher Oxytocin Levels Are Linked to Lower Blood Pressure and Heart Rate in Premenopausal Women," *Biological Psychology* 69, no. 1 (2005): 5–21, doi.org/10.1016/j.biopsycho.2004.11.002.

23. J. H. Chen, T. M. Gill, and H. G. Prigerson, "Health Behaviors Associated with Better Quality of Life for Older Bereaved Persons,"

Journal of Palliative Medicine 8, no. 1 (2005): 96–106, doi.org/10.1089 /jpm.2005.8.96.

24. S. S. Rubin, "The Two-Track Model of Bereavement: Overview, Retrospect, and Prospect," *Death Studies* 23, no. 8 (1999): 681–714, doi.org /10.1080/074811899200731.

25. Erich Fromm, *Escape from Freedom* (Farrar & Rinehart, 1941).

Chapter 9: Awareness, Attention, Intention, and Purpose

1. Yalom and Yalom, *A Matter of Life and Death*, 162.

2. R. L. Tate et al., "The Single-Case Reporting Guideline In BEhavioural Interventions (SCRIBE) 2016 Statement," *Physical Therapy* 96, no. 7 (2016): E1–10, doi.org/10.2522/ptj.2016.96.7.e1.

3. Clare Menzel, "Abbie's Way," Flathead Beacon, March 26, 2021, flatheadbeacon.com/2021/03/26/abbies-way/.

Chapter 10: Love and Social Connection

1. Sharon Salzberg, *Real Love: The Art of Mindful Connection* (Flatiron Books, 2017).

2. "Loving-Kindness Meditation with Sharon Salzberg," Mindful, accessed July 17, 2024, www.mindful.org/loving-kindness-meditation-with -sharon-salzberg/.

3. Erich Fromm, *The Art of Loving* (Harper & Row, 1956), 46, ia801309.

Index

About the Author

Mary-Frances O'Connor, PhD, is a professor of psychology at the University of Arizona, where she directs the Grief, Loss, and Social Stress (GLASS) Lab, investigating the effects of grief on the brain and the body. She is the author of *The Grieving Brain: The Surprising Science of How We Learn from Love and Loss*. O'Connor holds a PhD in clinical psychology from the University of Arizona and completed a postdoctoral fellowship in psychoneuroimmunology at the UCLA Semel Institute for Neuroscience and Human Behavior. She has authored research papers published in a wide range of peer-reviewed journals, from the *American Journal of Psychiatry* to *Psychological Science*. A fellow of the Association for Psychological Science since 2019, O'Connor also won the Patricia R. Barchas

Award in Sociophysiology from the American Psychosomatic Society in 2023. O'Connor's work has been discussed in the *New York Times*, the *Guardian,* the *Washington Post*, and *Scientific American*. She traded the snow of Montana, where she grew up, for the monsoons of Arizona, where she now lives.